Loss Control Auditing

Loss Control Auditing: A Guide for Conducting Fire, Safety, and Security Audits is a one-stop resource for both developing and executing a loss control audit program. This fully updated second edition addresses loss control auditing from the perspectives of workplace safety, physical security, and fire risks. It focuses on the three core areas of an audit: documentation review, physical inspection, and employee interviews, and presents a three-phase model – pre-audit, audit, and post-audit activities – which can be used for all three core areas.

This new edition benefits from the addition of auditing and system measurement material as promulgated in ISO 45001 and ANSI/ASSP Z10 standards and the Occupational Safety and Health Administration's Recommended Practices for Safety and Health Programs. It offers an expanded discussion of the application of auditing to the field of emergency management and new text explaining how leading and lagging measures can be used in the auditing process during assessment as well as in the post-audit evaluation. Subsidiary organizations and their integration into the auditing process, such as the areas of contractor management and temporary worker safety are covered in detail. The book discusses the integration of qualitative and quantitative measures in an effort to arrive at a more holistic scoring mechanism to assess organizational performance. In all, the depth of material presented in this thorough book showcases how to develop and execute a loss control management system audit program to a high quality.

An ideal read for industry professionals, students, and postgraduates in the fields of fire service, loss prevention, and safety management.

Prof. E. Scott 'Scotty' Dunlap is a Professor in the School of Safety, Security, and Emergency Management at Eastern Kentucky University, USA. He had a 15-year career in practice prior to entering academia, including safety roles in residential mental health, warehousing and distribution, and agriculture settings. His current research focus on industry leader development and involvement in workplace safety and health and safety management system assessment.

Occupational Safety & Health Guide Series

Series Editor: Thomas D. Schneid, Eastern Kentucky University, Richmond, Kentucky

Loss Control Auditing
A Guide for Conducting Fire, Safety, and Security Audits

Second Edition

E. Scott Dunlap

CRC Press
Taylor & Francis Group
Boca Raton London New York

CRC Press is an imprint of the
Taylor & Francis Group, an **informa** business

Designed cover image: © Shutterstock

Second edition published 2024
by CRC Press
2385 NW Executive Center Drive, Suite 320, Boca Raton FL 33431

and by CRC Press
4 Park Square, Milton Park, Abingdon, Oxon, OX14 4RN

CRC Press is an imprint of Taylor & Francis Group, LLC

© 2024 E. Scott Dunlap

First edition published by CRC Press 2011

Library of Congress Cataloging-in-Publication Data
Names: Dunlap, E. Scott (Erik Scott), author.
Title: Loss control auditing : a guide for conducting fire, safety, and security audits / E. Scott Dunlap.
Description: Second edition. | Boca Raton, FL : CRC Press, 2024. |
Series: Occupational safety & health guide series | Includes bibliographical references and index.
Identifiers: LCCN 2023036476 (print) | LCCN 2023036477 (ebook) |
ISBN 9781032442938 (hbk) | ISBN 9781032436852 (pbk) |
ISBN 9781003371465 (ebk)
Subjects: LCSH: Fire prevention--Inspection. | Industrial buildings--Fires and fire prevention. | Loss control. | Industrial safety.
Classification: LCC TH9176 .D86 2024 (print) | LCC TH9176 (ebook) |
DDC 658.4/7--dc23/eng/20230812
LC record available at https://lccn.loc.gov/2023036476
LC ebook record available at https://lccn.loc.gov/2023036477

ISBN: 978-1-032-44293-8 (hbk)
ISBN: 978-1-032-43685-2 (pbk)
ISBN: 978-1-003-37146-5 (ebk)

DOI: 10.1201/9781003371465

Typeset in Times LT Std
by KnowledgeWorks Global Ltd.

Contents

PART I   *Evolution and Benefits of Auditing*

PART II   *Defining a Management System*

PART III   *Audit Components*

PART IV   Audit Phases

PART V   Management System Audit Development

PART VI   Audit Pathways

PART VII Audit Program Opportunities

Preface

A fundamental principle of quality management is that things which are measured receive organizational attention. A measurement of activity provides a concise tool that can be presented to management to communicate how well that activity is performed. Auditing fire, safety, and security programs is a critical avenue through which information can be derived from pertinent management systems and presented in a way that allows organizational leadership to know how those functions are performing.

Auditing fire, safety, and security programs also provides a comprehensive metric that allows understanding of how the complete management system functions. When properly designed, an audit will consider all aspects and variables that impact the function rather than focusing on a single component. For example, a metric that is used within the context of many safety management systems is an injury rate. This rate reflects the number of injuries experienced within an organization per a specified number of employees. As an injury rate becomes lower, it may be assumed that the safety management system is improving. This assumption can only be proven true if a comprehensive audit of the safety management system is conducted. Such an audit might include a review of written compliance program content, employee training records, a physical inspection of the work environment, and interviewing a sample of employees. Rather than utilizing only a focused metric to determine the level of performance of a program, such as an injury rate, a comprehensive audit will review all aspects of a program to determine on a detailed level those things which are functioning well and those which have opportunity for improvement.

This book will provide information on the audit process, creating an audit document, creating an audit program, fire auditing, workplace safety auditing, security auditing, and audit scoring. This book will be beneficial in the development of a comprehensive and effective audit program.

Words such as "may" and "might" are used frequently throughout this book. These words are used because there is no predictable formula for what will work in any given organization. Auditing is not an exact science. Experimentation may be required to determine what will be successful in a specific environment. This book does set forth a methodology that is true in its basic form, but work will need to be done to determine the best way to implement the principles within a given organization based on variables such as organizational culture and policies, auditor skill level, and audit program budget.

Introduction

Auditing is a tool that can be utilized within any organization to identify where things are going well in addition to identifying where opportunities for improvement exist in a management system. A high-ranking executive has scheduled a visit to your facility and you want to ensure that everything is in good shape for their arrival. You conduct a review of the facility to ensure housekeeping standards are maintained and that processes are functioning properly. This is a basic form of an audit.

An audit is simply the measurement of existing performance against an established standard. Throughout this book we will review how auditing can be implemented in the areas of fire, workplace safety, and security. This will be accomplished by first discussing the evolution and benefits of management system assessment in Chapters 1–4 in Part I.

Part II of the book will explore management system standards and the components of a management system in Chapters 5–10. ANSI/ASSP Z10 and ISO 45001 will be addressed. Though these consensus standards address Occupational Safety and Health Management Systems, the concepts are transferable to the fields of fire and security.

Part III of the book examines the three primary components of an audit. Chapters 11–13 address the three core audit activities, which are the documentation review, conducting a facility inspection, and conducting employee interviews.

Part IV of the book presents the three primary phases of an audit program. Chapters 14–16 address these phases as being pre-audit, audit, and post-audit activity.

Part V of the book addresses the development of a comprehensive audit program. Chapter 17 provides detail in creating an audit document using Microsoft Excel. Microsoft Excel is used as the tool to develop an audit document due to its accessibility as a part of Microsoft Office and the mechanics learned from the material presented can be applied if auditing software or applications are used in an organization. Chapters 18–21 address the systemic auditing issues of audit scoring, auditor selection and training, audit logistics, and audit frequency. Chapter 22 presents the components of a written audit program that can be used to guide the auditing process in an organization.

Part VI of the book addresses the primary pathways in which audits can be conducted. Chapters 23–26 focus on the specific issues of conducting audits in the areas of workplace safety, security, fire, and creating mini audits.

Part VII of the book concludes with a discussion of opportunities related to conducting the audit. Chapter 27 discusses how the audit can be used as a training tool as the audit is being conducted. Literature from effective management techniques to Occupational Safety and Health Administration material emphasize the need to involve employees. This issue is addressed in Chapter 28. Chapter 29 navigates challenges that might be faced in the audit process. Chapter 30 closes with a discussion of auditing implications in relationships that exist between parent companies and subsidiary organizations.

Each chapter concludes with a case study to assist in applying the information covered in that chapter to fire, safety, and security issues. Questions are also provided at the end of each chapter to stimulate critical thinking on the topics that have been presented.

Part I

Evolution and Benefits
of Auditing

1 Organizational Levels and Auditing

Organizations implement numerous systems as they work to achieve their desired aims. For example, manufacturing facilities implement human resources, finance and accounting, and operations systems. Fire departments implement preventive maintenance, training, and testing systems. Hospitals implement quality control, patient care, and insurance systems. Auditing is a tool that is utilized to assess the performance of these systems. Auditing can present itself in numerous manifestations based on the system being audited and the organizational context of the audit. The variation in application of auditing is the basis for a focus on tools and principles throughout this text.

Auditing in safety, security, and fire is important due to the impact of associated systems on organizations. The integrity of systems used to manage these disciplines has a direct impact on the protection of people and property. Many organizations communicate that "their employees are their most valuable asset," which addresses the need to protect workers and managers through the implementation of appropriate management systems. Failure to protect workers can result in a spectrum of direct and indirect loss, such as:

- Workers' compensation costs (lost wage coverage, medical expenses, and claim maintenance costs)
- Loss of quality of life to the worker
- Emotional, physical, and financial impacts on the worker's family
- Emotional impact on uninjured co-workers and managers
- Tarnishing the brand of the organization
- Loss of production volume through replacing an injured experienced worker with a less experienced worker
- Loss of production quality through replacing an injured experienced worker with a less experienced worker
- Production time lost due the engagement of the Employee Assistance Program

Property must also be protected to ensure the ability to maintain organizational operations. Incidents that cause property damage can range in a minor repair to a facility being non-operational for a period of time. Property damage can result in a spectrum of loss, such as:

- Cost of repairing or replacing damaged equipment
- Tarnishing the brand of the organization
- Short-term loss of business during non-operational periods

DOI: 10.1201/9781003371465-2

- Long-term loss of customers who found other suppliers during non-operational periods
- Loss of employees who had to seek employment elsewhere due to facility shutdown
- Increased insurance premiums due to claims made to accomplish repairs
- Injured workers
- Fatally wounded workers

The potential for these losses to occur creates the need to conduct audits to measure systems that impact the protection of people and property in organizations. Creating and implementing appropriate systems is an important first step in the process. Auditing these systems is a critical second step to ensure they are operating at their fullest potential.

THE IMPORTANCE OF MEASUREMENT

Measurement provides the ability to assess the degree to which elements of a system or the system as a whole are functioning. Types of measurements include:

- Lagging measurements
- Leading measurements

Lagging measures provide a measurement of historical performance. These measurements assess things that have already occurred. Lagging measures are used routinely in organizations, such as measuring production volume and production quality. These measurements provide the ability to examine the history of production volume and quality over a specific period of time in the past. This information can then be used to assess why desired performance was not achieved and make improvements for future activity. Lagging measurements have long been utilized in industry and government to assess performance, which makes their translation into safety, security, and fire a natural progression. Examples of lagging measurements in the field of occupational safety include metrics such as:

- Number of injuries
- Number of fatalities
- Injury rates (recordable injuries, restricted work injuries, and lost time injuries)
- Fatality rate
- Workers' compensation costs
- Number of property damage incidents
- Property damage costs

Examples of lagging measurements in the field of security include metrics such as:

- Incidents of theft
- Workplace violence incidents

- Number of physical security system deficiencies
- Number of security system breaches
- Cost of security-related loss

Examples of lagging measurements in the field of fire include metrics such as:

- Number of structure fire responses
- Number of auto accident responses
- Number of emergency medical services (EMS) responses
- Occupational safety losses mentioned previously
- Security losses mentioned previously

Lagging measures are typically presented in the form of a volume or a rate. Some lagging measurements might be best presented in the form of volume. For example, workers' compensation loss can be presented in the form of volume of dollars lost. Where possible, a rate of occurrence is preferred over a volume of occurrence. The volume of occurrence does provide useful data but has a limitation of not considering changes in the amount of work performed in one period compared to another. For example, the formula for calculating an injury rate is

$$\text{Number of Injuries} \times 200,000/\text{Actual Hours Worked}$$

The number 200,000 represents a constant of 100 employees working 40 hours per week for 50 weeks in a year, assuming two weeks of vacation. The rate produced communicates the number of incidents per 100 workers. It might be found that a facility experienced the following in a fiscal year of operation:

$$8 \text{ injuries} \times 200,000/452,032 = 3.5$$

The following fiscal year resulted in:

$$11 \text{ injuries} \times 200,000/645,895 = 3.4$$

When examining the volume of injuries, it appears the facility regressed in performance due to an increase of 3 injuries, from 8 injuries in the first year to 11 injuries in the second year. However, examining the rate presents a different result in that improvement is observed by a slight reduction of injury rate from 3.5 in the first year to 3.4 in the second year. The rate measurement provides a more reliable metric in that it considers the fluctuation of hours worked.

Examining lagging measures that address historic performance provides the opportunity to understand where problems in the system occurred. The causes of these problems can be evaluated, and processes put in place to reduce loss in the future. For example, the lagging measure of an injury rate for the most recent business quarter might indicate an increase over the previous quarter. A deep examination of incidents might reveal that many incidents were associated with soft tissue injuries. This awareness can result in an increased focus on ergonomic program improvements to prevent future incidents.

Leading measures provide the opportunity to proactively observe and measure activity that should predict the prevention of loss to people and property in the future. The aim is to collect data on existing activity that can indicate how well the system is performing. Corrections can be made based on collected data with the intent of preventing future incidents. For example, a behavioral observation form can be created that assesses the behaviors a security officer must execute while processing employees and visitors at a point of entry. The behaviors that must be executed are:

- Confirm right to enter (identification badge for employees is present and ensure visitors are on an approved list)
- Examine the contents of all containers carried into the facility, such as purses, backpacks, and toolboxes
- Process each person through a metal detector

A form can be created with three columns. The first column lists each of the three behaviors. The second column provides cells where a tick mark can be entered when each behavior was accomplished properly. The third column provides cells where a tick mark can be entered when each behavior was accomplished improperly. The tick marks can be tallied and converted into a percentage of proper behavior. For example, the following might have been observed after the processing of 20 individuals into the facility:

- *Confirm right to enter*: 18 correctly confirmed; 2 incorrectly confirmed (18 correct behaviors/20 total behaviors = 90%)
- *Check containers*: 15 correctly checked; 5 incorrectly checked (15 correct behaviors/20 total behaviors = 75%)
- *Metal detector processing*: 20 correctly processed; 0 incorrectly processed (20 correct behaviors/20 total behaviors = 100%)

These measurements present a great deal of opportunity to reinforce the need and process to check containers with security officers due to that behavior occurring at 75% correct. Confirming the right to enter also provides the opportunity for reinforcement due to the behavior occurring at 90% correct. Metal detector processing indicates the level of behavior desired at 100% correct.

This type of leading measurement activity can be viewed as a form of an audit in that it is an assessment of a part of the system with the output of a metric that can be used to influence performance improvement. Audits can be used as a leading measure to assess a part of a system or a system as a whole to identify potentials for improvement prior to an incident occurring.

INSPECTIONS VERSUS AUDITS

Inspections have long been used as a tool to identify potential problems and address relevant issues before an incident occurs. For example, fire inspections have been used to identify fire hazards with the goal of correcting problems before a fire occurs. Similarly, forklift pre-use inspections are conducted to ensure the piece of equipment is safe to operate before using it to engage in work.

Findings from inspections do not typically produce a metric that is useful to measure system performance. They are simply designed to find and correct problems. In the example of forklift pre-use inspections, a large volume of pre-use inspections might be performed, and the system might be perceived as working well when:

- All deficiencies are reported timely in a maintenance work order process
- Repairs are made in a timely fashion
- Equipment utilization is maintained at a high level by timely repairing and placing the forklift back in service

However, an audit process that produces metrics might provide added value to the organization. An analysis of forklift inspections might reveal a metric of 63% of maintenance work orders being related to replacing damaged tires. Further exploration of this issue reveals three areas in the facility with damaged concrete, which causes the damage to forklift tires. Investment in repairing the damaged concrete prevents future damage to forklift tires. Creating a robust audit process allows the opportunity to explore the root causes of system failures and implement corrective measures prior to the occurrence of incidents.

Throughout this text, an inspection will be interpreted as activity that simply generates items to be corrected. An audit elevates the level of useful information by providing quantitative or qualitative measurement that can be used to monitor system performance.

APPLICATION TO LEVELS IN ORGANIZATIONS

Auditing can be a useful activity to measure performance within and among various operational groups. Organizational levels can be useful in implementing measurements due to different members of leadership who will correlate to each level. Organizational levels can include the following:

- Departments
- Shifts
- Facilities
- Regions

Auditing and producing the measurement outcomes of the process can be useful in benchmarking performance at the departmental level. For example, behavioral auditing among forklift operators in different departments can serve as an internal benchmark. It might be found that forklift operation is performed at a level of

84% safe in the shipping department compared to 97% safe in the receiving department after observation of behaviors such as pre-use inspection, seatbelt use, horn use, operation speed, and interaction with pedestrians. This benchmark can serve as motivation for the shipping department to improve as well as clarifying which behaviors might hold the most opportunity for improvement.

Auditing and producing the measurement outcomes at the shift level can be useful in a similar fashion as to that of departments. The primary difference is that shift managers can drive change across multiple departments. Audit results can be communicated to shift managers and can clarify system performance across shifts. The benefit of producing audit results at the shift level is that cohesive solutions to problems can be implemented uniformly across departments on shifts and motivation for improvement can occur on a shift that is deemed as lesser performing.

System audits can be used to benchmark performance among multiple facilities in an organization. Such comparisons will allow insight into how well upper management at each facility is supporting the implementation of programs. For example, an audit can be developed and implemented to assess safety and health management system performance that aims to protect workers from injury and fatality and facilities from property damage. Audit outcome measurements of system components and the system as a whole communicate how each facility is performing in protecting people and property. Facility management can benchmark how well they are performing compared to other facilities in the organization and examine specific opportunities for improvement based on metrics produced by the safety and health management system audit.

Audit outcome measures can be communicated on regional levels. Though location audit outcomes can be communicated, communicating audit results on a regional level can assist in engaging regional managers in the process of driving system performance improvement among facilities within their scope of responsibility.

AUDIT TOOLS

Audit tools vary in nature with a spectrum of tools utilized across various sizes of organizations. Though technology has resulted in an evolution of audit tools, they are not necessarily selected for use. The challenge is to identify the type of tool that will serve the purposes of an organization. A scope of options of tools can include:

- Hard copy documents
- Microsoft Excel spreadsheets
- Web-based or app-based tools

Hard copy documents might be considered as being outdated but are still widely used for a number of reasons. First, hard copy audit documents do not rely on technology to function properly. Hard copies can be used to collect data through the simple use of pen and paper. The data can then be entered into a spreadsheet or database for further analysis. Second, physically handling a hard copy of an audit has the perceived benefit of closely understanding what is occurring by observing behavior and writing responses and notes on an audit form. Third, hard copy processes might come with a lower financial investment in the process.

The use of Microsoft Excel is an initial step in utilizing technology in auditing. Microsoft Excel provides a framework to develop a robust and flexible management system audit that can perform complex calculations of varying scoring methodologies resulting in quantitative outcomes. These outcomes can be used to assess topical components of the system and comprehensive measurements of system performance. Microsoft Excel files can be used on a tablet, which makes the format versatile in use while walking through a facility or working in an office to enter assessment information. This text will utilize Microsoft Excel as a framework in developing a management system audit due to:

- Wide accessibility on most computers as a standard component of Microsoft Office
- The ability to immediately use a created document in conducting an audit
- Immediate ability to edit and evolve the audit document as needed over time

Understanding the mechanics of developing a management system audit in Microsoft Excel will also be of use in understanding designs behind various web-based or app-based auditing tools. This understanding will assist in interacting with product designers to customize audit tools to meet the needs of a given organization.

Web-based or app-based auditing tools present the ability to leverage technology through pre-designed auditing tools that can be customized to meet the needs of an organization. Limitations of various products will need to be investigated as certain audit products might not be useful in creating a tool in design or outcome measurements that meets all the needs of an organization. Technology support and hardware will also need to be considered in the implementation process. For example, tablets might need to be purchased to support the use of a given app-based audit tool and immediate technology support might be needed if the audit tool fails to function while an audit is being conducted.

CASE STUDIES

FIRE

Bill has recently been promoted as chief of his fire department. City leadership is concerned with a number of incidents in the community that have made the evening news to include fire and auto accidents. Bill is tasked with identifying how his fire prevention education department can be used to reduce loss that is occurring.

SAFETY

Jean is the safety manager at a multi-site manufacturing organization. The Occupational Safety and Health Administration (OSHA) inspected one of her facilities and was found to be lacking in safety and health management system development. Upper management is concerned due to the potentially negative impact poor occupational safety performance can have on the brand of the organization. They are impressing upon her the need to improve performance.

SECURITY

Jose is the area manager over distribution for an international apparel company. He forgot his identification badge when coming to work, but security allowed him to enter the facility where his office was located. He was concerned about the behavior of security officers because they did not call anyone to ensure he was still employed with the organization, regardless of his position. He had a conversation with the security manager about the deficiency.

- How can auditing be used to impact performance improvement?
- What audit process might be most useful in addressing the issues?

EXERCISES

For the following questions, identify a single facility environment in which you would like to situate your responses and answer each question accordingly. Answer each question in the context of a fire, safety, or security audit.

1. Why should auditing be implemented in the organization?
2. What lagging measures are used to measure system performance?
3. What leading measures are used to measure system performance?
4. How can inspections be used and what are their limitations?
5. How can auditing be used to establish outcome measurements of the management system?
6. How can auditing be used to most effectively foster performance improvement at different levels within the organization?
7. What audit tool is most appropriate given the context and resources of the organization?

2 Inspections

Inspections have long been used as a tool to identify deficiencies in a given area. Inspections typically involve the use of a standardized form to prompt the inspector to observe certain items on a checklist. The results of an inspection do not usually include a quantitative measurement but rather a list of deficiencies that need to be addressed.

Following are examples of ways in which inspections have been utilized. The information included in this chapter should not be considered comprehensive but to serve as examples of the use of inspections in various workplaces. Exploration and application can be made within unique organizations to identify how inspections have been and can be used to assess performance.

INFORMAL INSPECTIONS

Informal inspections often occur in the form of a "tour" or "walk through" of an area. Areas might be confined to a department or specific production area to maintain focus on what is being observed. A checklist is not necessarily used during this type of activity and is instead used as an informal assessment of the condition of the work area.

Personnel involved in an informal inspection can include one or a few of the following:

- Department Supervisor
- Shift Manager
- Plant Manager
- Safety Manager or Security Manager, depending on the intent of the informal inspection
- Workers

A Department Supervisor might routinely walk through their area of responsibility to ensure proper physical conditions are in place and work is performed as intended. A Shift Manager or Plant Manager can be present during an informal inspection to remain engaged in current issues that impact a given work area. This can be an opportunity to maintain open communication between line management and middle and upper management. Depending on the focus of the informal inspection, the designated manager responsible for that discipline can be present to have a timely dialogue on immediate issues faced by those in the work area. For example, the Safety Manager can accompany the Department Supervisor during an informal inspection and discuss unique safety issues and potential solutions as various areas of the department are visited. Workers can participate in informal inspections and offer their unique perspectives on how

DOI: 10.1201/9781003371465-3

work is performed. They can provide unique insights on recommendations for process improvements that can lead to heightened levels of production as well as safety and security.

HOUSEKEEPING INSPECTIONS

Housekeeping inspections can be performed by a Department Supervisor or trained employee. Housekeeping inspections address basic issues of exposure to risks of injury and fire. These inspections follow a prescribed document that prompts the inspector to explore specific housekeeping issues, which can include:

- Floor areas are free of debris
- Items are properly stored
- Waste is properly collected and disposed
- General orderliness of the work area

The degree of risk will play a significant role in the meaning of housekeeping inspections for workers and management. Loose paper on the floor of an office area might be perceived as a minor housekeeping problem. However, a worker can experience a simple fall from slipping on the piece of paper, which can result in:

- Extensive surgery to repair a broken elbow suffered in a fall
- Workers' compensation costs that include lost wages to the injured worker while they are away from work healing from the surgery and medical costs for the surgery, follow-up appointments, and physical therapy
- Loss in production volume and quality from replacing the injured worker with a less experienced worker

Housekeeping can be of great importance in unique workplaces, such as in the grain industry. Grain elevators exist throughout the Midwest and are used to store grain originating from farmers, such as corn. The movement of grain throughout an elevator generates dust that can accumulate throughout the facility. Grain dust is an explosion hazard in elevators, which makes housekeeping activity and inspections an issue of paramount importance in that work environment.

FIRE INSPECTIONS

Fire inspections are used to assess conditions related to fire prevention and hazard control. These are documented inspections that can occur on a weekly, monthly, quarterly, or annual basis. The focus of these inspections is to ensure fire risks are properly managed through facility processes and systems. Multiple types of fire inspections can be used within a facility, to include:

- Fire hazard inspections
- Fire extinguisher inspections
- Fire system inspections

Fire hazard inspections are designed to specifically address fire risks that are unique to facility operations. These inspections are typically performed on a weekly basis due to the evolving nature of operations. Issues can surface on an agile basis that can affect fire safety. For example, excessive amounts of recyclable cardboard might accumulate in the facility, which could present substantial fire load. A weekly fire hazard inspection can be used to ensure issues such as this are managed in a timely manner. A fire hazard inspection can also be used to ensure facility fire suppression system components are properly managed, such as sprinkler heads remaining unobstructed as well as having open access to the fire pump and risers.

Fire extinguisher inspections are performed monthly. This inspection does not use a form, as is used with other types of inspections. Fire extinguisher inspections are documented using a tag affixed to each fire extinguisher. The intent of this inspection is to ensure each fire extinguisher is in good physical condition and is properly charged. The inspector will initial and date the tag if the extinguisher is in good condition and properly charged. A diagram of the facility with the location of each fire extinguisher identified will help to ensure all fire extinguishers are in place and inspected each month.

Fire system inspections are conducted to ensure the integrity and operational ability of components of fire suppression systems. These inspections might require the engagement of contractors or property insurance personnel who are subject-matter experts in testing these systems. In-house maintenance personnel can also be trained to perform the appropriate inspections. Fire system inspections can include:

- Operation of post-indicator valves
- Fire pump operation and flow capacity
- Trip testing fire system risers

A Safety Manager can perform a post-indicator valve inspection. This inspection simply involves ensuring post-indicator valves are locked in the "open" position and closing and reopening the valve to ensure it functions properly. Fire pump inspections might require the use of multiple resources:

- A Safety Manager should be able to perform a basic fire pump operation test to ensure the fire pump activates by running it for a period of time.
- Technical resources in a properly trained maintenance department, contractor, or insurance personnel might be needed to perform flow tests to ensure the pump can flow an appropriate amount of water.
- Technical resources in a properly trained maintenance department, contractor, or insurance personnel might be needed to perform fire system riser trip testing to ensure risers function properly and are effectively reset.

SAFETY AND SECURITY INSPECTIONS

Safety inspections are typically performed monthly. Safety inspections are designed to observe physical condition issues that could lead to injury, illness, or fatality among workers as well as property damage to the facility. These inspections are more robust than a targeted housekeeping or fire inspection, which can require more time to accomplish. Safety inspections can include the observation of the following:

- Machine guarding
- Chemical storage and labeling
- Availability of Safety Data Sheets
- Unobstructed means of egress in the event of an emergency
- Presence of railing on elevated working surfaces
- Condition of tools and equipment used by maintenance personnel
- Condition of ladders and steps used to access elevated areas
- Condition of working surfaces
- Performance of emergency lighting
- Performance of alarm systems
- Clear access to electrical panels
- Safe storage of products, tools, and equipment
- Proper condition and availability of personal protective equipment (PPE)
- Presence and appropriateness of signs and tags

Security inspections are typically performed monthly. Security inspections are designed to observe physical condition issues that could lead to the loss of product or unauthorized access to certain areas within the facility. Security inspections can include the observation of the following:

- Proper function of access control components, such as metal detectors, gates, or badge access systems
- Performance of alarm systems
- Integrity of perimeter fencing
- Effective placement of security officers throughout the facility

PRE-USE INSPECTIONS

Pre-use inspections follow a specified checklist of items to ensure the targeted piece of equipment is safe to use. These inspections can be conducted in a short amount of time and address specific operational issues. The operator of the equipment to be inspected will be trained to properly conduct the inspections. Examples of pre-use inspections include:

- Powered industrial equipment
- Vehicles operated on public roads (those regulated by the Department of Transportation as well as fleet vehicles)
- Hot Work Equipment
- Fall Protection Equipment

Powered industrial equipment is routinely used throughout distribution centers and manufacturing facilities to move pallets of product. This equipment includes things such as forklifts, stock pickers, reach lifts, and pallet jacks. Organizations may incorporate other pieces of equipment into this process, such as wranglers, bob-cats, riding lawn mowers, and maintenance carts. The Department of Transportation (DOT) mandates pre-use inspections that must be performed on commercial motor vehicles. These pre-use inspections are managed under an organization's transportation safety program and can be expanded to be used for fleet vehicles that are not under the scope of the DOT. Pre-use inspections can be conducted on hot work equipment, such as welders and grinders, to ensure they are safe to use prior to working on the production floor, where fire hazards exist. Pre-use inspections can be conducted on fall protection equipment to ensure all equipment components are in safe operational condition, including harnesses, connection devices, and anchorage points.

EVOLUTION OF INSPECTION TO INCLUDE OBSERVATION

Inspections tend to be used to examine static items, as addressed in the discussion to this point. However, the term "inspection" has been expanded to address the observation of activity. The inspection of behaviors can help ensure programs and procedures are implemented as intended.

OSHA has addressed the inspection of behavior in the lockout/tagout standard. 29 CFR 1910.147(c)(6)(i) states, "The employer shall conduct a periodic inspection of the energy control procedure at least annually to ensure that the procedure and the requirements of this standard are being followed." Though the discussion of inspections to this point has focused on observing static items, the requirement of OSHA in this regulation introduces the use of observing worker behavior as a form of inspection. To effectively conduct a lockout/tagout inspection, the employer must observe worker behavior to ensure procedures are being followed. Planning must be used to record the occurrence of observations while lockout/tagout is actively used during work processes. Lockout/tagout might not be used on a planned basis due to the nature of the work being performed. For example, lockout/tagout is needed by maintenance personnel when equipment must undergo repair or maintenance due to ongoing use. Lockout/tagout might also be needed on a random basis when workers must clear processing equipment jams. These events occur on a non-routine basis, which requires a mechanism to be in place to alert those who might need to record an inspection of the lockout/tagout process as it is occurring.

Behavior-based observations are a growing component of management system assessments. Like lockout/tagout, planned observations can be made of workplace activity to ensure it is being performed as designed. These inspections include the creation of a document to record worker behavior.

- Column to list behaviors needed to properly carry out the task
- Column to record the number of times each behavior is carried out properly
- Column to record the number of times each behavior is carried out improperly

For example, in the realm of security, a behavior-based inspection might be conducted to assess how well access control procedures are being implemented at the point of entry into a building. The identified list of behaviors that must be carried out by security officers includes:

- Confirm right to enter by examining identification badges
- Process individuals through a metal detector
- Examine the contents of bags brought into the facility

A form can be created to assess whether each of these behaviors is performed properly or improperly by security officers as numerous workers enter the facility.

ADDRESSING INSPECTION FINDINGS

A system must be in place to address findings from the various inspections that are used within an organization. The corrective action system will help to ensure continuous progress is made in managing the system, supported by inspections. Three general categories of correcting findings include:

- Immediate correction
- Maintenance work order system
- Submit formal request for funding larger repairs

Deficiencies found in an inspection might be addressed immediately. These deficiencies can be addressed while the inspection is underway or within a timeframe that is immediate to the inspection. For example, a fire inspection might reveal that a product is stored too close to a sprinkler head. If the product is allowed to remain in place, the water plume of an activated sprinkler head would be obstructed, preventing water from reaching a fire below. In this case, maintenance or production personnel can be immediately contacted to remove and relocate the product so that the sprinkler head can achieve the level of protection for which it was designed.

Inspection deficiencies might require the use of a maintenance work order. The maintenance work order system provides a systematic method to correct problems found in the facility. For example, a safety inspection might reveal that emergency lighting at an exit is not working properly. A maintenance work order can be generated that will prompt maintenance personnel to correct the problem and record the completion of the activity.

Some inspection findings might result in the need for a much larger expenditure to address than can be accommodated in a maintenance work order. For example, a safety inspection might find that maintenance workers must routinely access and service equipment that exposes them to a fall hazard. The solution is to build work platforms and handrails to fully protect them against exposure to a fall. This solution would require capital investment due to the cost involved in designing and building appropriate work surfaces and railings. Such a request for funding would need to be submitted and processed through established procedures.

CASE STUDIES

FIRE

Shannon has worked tirelessly to develop the plant fire protection program. A recent visit from the Fire Marshal resulted in citations related to violations of the Life Safety Code. Exit routes were found to be problematic in addition to the fire alarm system not functioning properly.

SAFETY

Susan has worked with her maintenance staff to increase lockout/tagout performance to prevent injury. She has worked to ensure all maintenance workers are trained in lockout/tagout and the appropriate equipment are at their disposal. She has engaged maintenance personnel to ensure each piece of equipment has a lockout/tagout procedure to control hazardous energy.

SECURITY

Gene, the plant manager, was conducting an informal inspection with a department supervisor in the shipping department and noticed a contractor was working on a damaged dock door. He was not familiar with the company name on the uniform of the contractor and was concerned about their presence in the facility.

- How can inspections be used to improve organizational performance?
- What specific inspection document can be created and implemented to address the issue?

EXERCISES

For the following questions, identify a single facility environment in which you would like to situate your responses and answer each question accordingly. Answer each question in the context of a fire, safety, or security audit.

1. How can informal inspections be integrated into operations?
2. How can housekeeping inspections be used to increase the level of performance?
3. What types of fire inspections should be implemented in operations?
4. How can robust monthly inspections be implemented to fully assess system issues?
5. What pre-use inspections should be used for various equipment?
6. How can behavioral observations be used to enhance inspection processes within the organization?
7. What processes should be in place to address the findings from inspections?

3 Compliance Audits

Compliance audits greatly expand the scope of assessment beyond what is addressed in inspections. Where inspections focus on a single topic, such as housekeeping, compliance audits focus on the scope of activity that occurs to achieve compliance with regulatory requirements.

REGULATIONS

Compliance audits are driven by compliance with local, state, and federal regulations. For example:

- Fire prevention regulations might apply at the county level that are managed through the governing fire department.
- Fire prevention regulations might apply at the state level that are managed through the office of the state fire marshal.
- Fire prevention regulations might apply at the federal level through the Occupational Safety and Health Administration (OSHA) and the National Fire Prevention Association (NFPA).

It is the responsibility of each employer to determine the scope of regulations that apply to a place of business and ensure compliance. Assurance of compliance can be accomplished by conducting a compliance audit.

Professionals in the fields of occupational safety, security, and fire must explore regulations at the local, state, and federal levels to determine whether they apply to their place of business. This can be accomplished by:

- Contacting or exploring the websites of applicable municipal and county offices that manage regulatory requirements on a local level
- Contacting or exploring the websites of applicable state agencies that manage regulatory requirements on a state level
- Contacting or exploring the websites of applicable federal agencies that manage regulatory requirements on a national level

A large portion of compliance efforts derives from compliance with federal regulations and consensus standards. Some of the primary ones that affect occupational safety, security, and fire include:

- *Occupational Safety and Health Administration (OSHA)*: OSHA promulgates regulatory requirements that impact worker safety and property protection. Though the term "OSHA guidelines" is often used, it is erroneous in nature. The word "guideline" communicates that the information is

DOI: 10.1201/9781003714653-4

optional and used for guidance. OSHA regulations are not optional, and compliance must be achieved where regulations apply to an organization. For example, a manufacturing facility is under the category of "general industry," which means the OSHA standards in 29 CFR 1910 apply:

- 29 is the "title" of Labor.
- CFR stands for Code of Federal Regulations.
- 1910 is the "part" that refers to general industry occupational safety "standards" or "regulations."

OSHA has generated and maintains specific bodies of regulations based on industry sectors:

- General Industry
- Construction
- Maritime
- Agriculture

- *Department of Transportation (DOT)*: DOT promulgates regulatory requirements that impact the transportation of both hazardous and non-hazardous material. Regulations address transportation through various modes, including public roads, rail, and water. Primary regulations that might affect many employers include:
  - 49 CFR, Subtitle B, Chapter 1 – Transportation of Hazardous Materials addresses the hazardous material table, shipping papers, placarding, driver training, and emergency response planning.
  - 49 CFR, Subtitle B, Chapter 3 – Federal Motor Carrier Safety Regulations that address commercial drivers license (CDL), driver qualifications, emergency equipment, maintenance of accident registers and reports, and drug and alcohol testing programs.
- *Environmental Protection Agency (EPA)*: The EPA can apply as many occupational safety professionals might also be assigned environmental compliance under a larger scope as an Environment Safety and Health Manager. The EPA promulgates regulatory requirements that impact the protection of the natural environment, as the title of the agency communicates, from industrial operations. Examples of areas of focus include controlling emissions that are released into the air and preventing waste from entering bodies of water (streams, rivers, lakes, and oceans). Regulations that address the protection of the environment are found in 40 CFR.
- *Federal Emergency Management Agency (FEMA)*: FEMA promulgates regulatory requirements that impact disaster and emergency management on community and national levels. Regulations that address disaster and emergency management are found in 44 CFR. Primary regulations that affect disaster and emergency planning include the following from 44 CFR, Chapter 1:
  - General requirements
  - Insurance and hazard mitigation
  - Fire prevention and control
  - Disaster assistance
  - Fire assistance
  - Preparedness

- *Department of Homeland Security (DHS)*: DHS promulgates regulatory requirements designed to provide national security from external threats and was established because of the attacks of 9/11. Primary regulations that can affect organizations when developing a security program include:
  - *6 CFR*: domestic security
  - *8 CFR*: aliens and nationality
- *National Fire Protection Association (NFPA)*: The NFPA promulgates consensus standards that affect fire protection and fire operations. Consensus standards differ from regulatory standards developed by federal agencies in that they are developed by committees of professionals in the field who reach a "consensus" on best practices and what should be in place given a specific topic. However, consensus standards can become a regulated issue in two situations:
  - A given NFPA standard can be "incorporated by reference" into a regulation. For example, OSHA incorporates many NFPA standards by reference in fire protection regulations.
  - NFPA standards can be adopted by the "authority having jurisdiction." For example, state or local government agencies can adopt NFPA standards as law and enforce them among organizations accordingly.
- *American National Standards Institute (ANSI)*: ANSI produces consensus standards that impact workplace safety. For example, ANSI Z10 is a consensus standard that addresses the development and implementation of an occupational safety and health management system. Like NFPA standards, ANSI standards can become a regulatory requirement if incorporated by a governmental agency.
- *International Standards Organization (ISO)*: ISO produces consensus standards that impact a spectrum of organizational operations. These standards are international in that compliance with and certification in various ISO standards communicates levels of organizational excellence in operations when conducting business with organizations in other countries. For example, ISO 45001 addresses the development and implementation of an occupational safety and health management system. Certification in ISO 45001 establishes the achievement of a benchmark of excellence in providing a safe and healthy work environment.

An understanding of federal regulations and consensus standards will assist professionals in further exploring potential regulatory requirements at the state and local levels. For example, though OSHA is a federal agency, certain states decided to operate "state plans" to govern workplace safety. Compliance with federal OSHA regulations may not guarantee compliance with a state plan's requirements, as state plans are required to meet or exceed federal requirements. In some cases, a state might exceed federal requirements, which will require a specific investigation into state regulatory requirements. Compliance audits can be designed and carried out once the applicable scope of regulations and consensus standards is identified. Compliance audits will typically consider the aspects of reviewing documents and physical conditions in the work environment.

DOCUMENTATION

A review of documentation provides the opportunity to assess the degree to which compliance with regulations is being set forth in programs, policies, and supporting documents. For example, a distribution center will have multiple forklifts operated throughout the facility. The presence of this equipment requires compliance with OSHA's powered industrial truck standard, 29 CFR 1910.178. The standard addresses the following:

- General requirements
- Classifications of powered industrial trucks
- Areas in which various powered industrial trucks can be operated
- Safety guards
- Fuel handling and storage
- Changing and charging storage batteries
- Lighting for operating areas
- Control of noxious gases and fumes
- Bridge plates
- Trucks and railroad cars
- Safe operation
- Training program implementation
- Training program content
- Refresher training and evaluation
- Avoidance of duplicative training
- Certification
- Truck operation
- Traveling
- Loading
- Operation of the truck
- Maintenance of industrial trucks

The organization will be responsible for developing written material that guides compliance with each aspect of the regulations. Written documentation typically includes:

- Policies
- Programs
- Training records
- Completed incident investigation forms
- Supporting documents
- Grants

Policies are typically short documents that provide brief and specific information on one or two pages. The information provided includes high-level lists and descriptions of what must be accomplished.

Written compliance programs include detail that addresses each aspect of a regulation for which it is designed to comply. Compliance programs can include the following as templates in their construction:

- *Purpose*: a brief statement of what is to be achieved by implementing the program
- *Scope*: individuals, departments, facilities, and activities covered by the program
- *Delineation of Responsibilities*: a list of job titles of each person who must engage in program implementation with an explanation of what individuals in each role must do
- *Procedures*: a detailed presentation of all activities that must occur to comply with the regulation
- *Equipment*: a list of equipment needed to implement the program
- *Training*: an overview of classroom and practical skills training that will be conducted and documentation used to record the occurrence of training, which can include a training log, exam, and certificate of completion
- *Program Review*: a brief statement of the cycle at which the program will be reviewed and changes made on an ongoing basis

Training documentation will be based on what is stipulated in the written program. Training logs are used to provide a list of workers and managers who attended a certain training session. Multiple-choice exams are often used as a form of comprehension verification. Certificates of completion are used as an agreement between the trainer and the trainee that a provided list of topics on the document was covered in the training session.

Investigation forms are used to demonstrate a history of thoroughly exploring why incidents occurred and identifying what can be done to prevent future incidents. Investigation forms can cover a broad scope of activity, including:

- First-aid incidents
- Injuries
- Illnesses
- Fatalities
- Security incidents
- Property damage incidents
- Fires
- Workplace violence incidents
- Disaster incidents

A broad scope of supporting documents can be presented. Each regulation will serve as a catalyst as to what supporting documents must be created and managed in the compliance process. For example, OSHA's confined space entry standard (29 CFR 1910.146) requires a review of confined spaces that pose a hazard prior to entry. This requirement is managed through issuing a document referred to as a confined space entry permit. Regulations must be explored to determine what activity must occur. Documents will then need to be created as needed to support compliance activities.

Grant documentation will need to be diligently produced and managed in cases where grant funding has been obtained to fund program improvement. For example, OSHA offers Susan Harwood training grant opportunities to assist in reaching unique populations with important training topics. Grants will typically require:

- Proposal that satisfies the scope of activity covered by the grant
- Accounting of funds distributed through the grant and subsequent expenditures that satisfy the information presented in the proposal
- Reports of activity accomplished through activity funded by the grant

PHYSICAL CONDITIONS

Compliance audits typically include a facility inspection to assess the physical conditions of the work environment in relation to activity driven by compliance programs. An assessment of the physical conditions will assist in determining how well what is presented in written compliance programs is being implemented in the workplace. For example, a manufacturing facility might contain places where workers must be protected through the implementation of a fall protection program. An assessment of physical conditions might include observing:

- Placement of handrails
- Condition and use of ladders
- Condition and use of harnesses and lanyards
- Identification of appropriate anchorage points
- Work processes near open dock doors

Each compliance program will need to be explored to determine the scope of physical condition issues that must be routinely reviewed to assure compliance.

REGULATORY INSPECTIONS

Compliance audits discussed to this point are typically conducted internally. The subject matter expert responsible for the scope of activity will conduct the compliance audit. For example, a Safety Manager employed by the organization will conduct a safety compliance audit. Beyond this internal activity, external regulatory inspections might occur. These compliance inspections are conducted by the agency responsible for administering the given body of regulations. For example, OSHA employs numerous Compliance Officers in both federal and state programs who are responsible for conducting compliance inspections at work sites. These inspections can be caused by the following:

- Randomly selected by the agency in ongoing proactive inspection processes
- Incident might occur that prompts an inspection.

- Whistleblower complaint that specifically addresses a compliance issue at a given facility

Employers must become aware of the catalysts for various governmental agency inspections and prepare for and manage them productively. Appendix C contains a checklist that can be used for such preparation.

CASE STUDIES

FIRE

Jill is the chief of the Smithville Fire Department. She has been working to lead the development of a compliance program that addresses the scope of departmental activity. She is concerned about both equipment and personal protection issues in the development of her compliance program.

SAFETY

Steve is the first Safety Manager to be hired in the manufacturing plant, where he has just accepted the position. Safety was managed in previous years through the assignment of responsibility to the Human Resources Manager. The Human Resources Manager admits to doing the best he could with what he knew about safety.

SECURITY

Eliot is the Security Manager for an electric company in the Midwest. His region predominantly covers rural areas of lightly populated communities. Though the electric company is considered as part of our country's critical infrastructure, little emphasis has been placed on developing a security program due to a perceived lack of risk given their rural coverage.

- What governmental agency or consensus standard-producing organization will need to be explored?
- How can the organization best prepare for and utilize compliance audits to assess performance?

EXERCISES

For the following questions, identify a single facility environment in which you would like to situate your responses and answer each question accordingly. Answer each question in the context of a fire, safety, or security compliance audit.

1. What governmental agencies will impact the development of your compliance program?

2. What consensus standards setting organizations will impact the development of your compliance program?
3. What forms of documentation will need to be in place to ensure compliance with various regulations?
4. How can an assessment of physical conditions assist in achieving compliance with applicable regulations or consensus standards?
5. What types of governmental regulations might you need to prepare for, and what are the catalysts for their occurrence?

4 Management System Audits

Assessments have evolved over time. Inspections have been used to identify deficiencies that must be corrected. Compliance audits have been used to determine the degree of compliance with local, state, and federal regulatory requirements. Management system audits are now used to assess the complete scope of regulated and non-regulated activity that must occur to ensure the system is functioning at its maximum capacity.

Management system audits have evolved as a more robust assessment due to the identified limitations of establishing regulatory compliance as an ultimate organizational goal. Regulatory compliance has been accepted as a component of organizational performance, but additional system components are needed to achieve an optimum level of performance. Regulatory compliance will result in a limited level of system performance. For example, OSHA is a macro-level organization that drives workplace safety requirements on a national level. The very nature of the agency leaves micro-level variables among organizations in question. Regulatory compliance alone will still result in losses due to additional organization-specific issues that must be considered. The gap in system performance between regulatory compliance and zero incidents occurring has been addressed through a comprehensive management system focus that considers:

- Leadership
- Worker participation
- Hazard identification
- Hazard control
- Education and training
- Management system assessment

The next part of this text will explore these components more fully, but here will be a general introduction to management system components as the current pinnacle evolution of auditing in safety, security, and fire.

LEADERSHIP

Leadership in supporting the implementation of a management system is necessary for the system to function at its greatest capacity. Organizational leaders do not necessarily need to become subject matter experts, but they do need to engage to the extent that demonstrates their commitment and support for the process. This can be accomplished through activities such as department supervisors including safety in a pre-shift meeting or the plant manager including safety on the agenda of a plant meeting.

DOI: 10.1201/9781003714653-5

Assessing leadership's engagement in the management system can be a challenging endeavor. Assessment can occur by reviewing meeting minutes or agendas to determine if the topic of the management system is present. For example, a plant meeting that contains safety on the agenda can be an indicator of leadership engaging in and demonstrating support for the safety management system. The operating budget could be reviewed to determine if safety is allocated appropriate funds to properly manage discipline in the organization. The goal is to identify ways in which leadership can demonstrate support and engagement in the management system and then determine methods to audit the activity.

WORKER PARTICIPATION

Similar to leadership engagement in the management system, ways for workers to participate in the management system should be identified and implemented. Worker participation is often viewed as a critical success factor for the management system. Workers perform tasks daily and are uniquely positioned to provide input as to how work can be accomplished more safely and share insights on fire and security issues. Tapping the knowledge and experience of workers can be an incredible asset in improving organizational performance.

Assessing worker participation can follow a similar process as assessing leadership's engagement in the management system. Workers can be engaged in a broad spectrum of activities, including participating in incident investigations, serving on a safety committee or first responder team, and providing suggestions for performance improvement. Worker participation should be viewed individually within various facilities. For example, workers in one plant might simply prefer not to be heavily engaged, while workers at another plant might want to be very engaged in the management system. Worker participation should not be forced, and the audit process should be flexible to measure engagement activities based on the unique dynamics of each facility.

HAZARD IDENTIFICATION

Hazard identification involves conducting a survey of the workplace to identify all hazards that exist that can result in worker injury or fatality and loss of or damage to company property. The process of hazard identification applies to all management systems. A safety hazard might be exposure to a fall from a height. A security hazard might exist at a point where the perimeter of the facility is not secure. A fire hazard might be the improper storage of flammable material. Regardless of the management system, hazards must be identified. Two tools are typically used to identify hazards:

- Job Hazard Analysis
- Risk Assessment

A job hazard analysis is used to break down a job into its various steps. Each step is then analyzed to determine what can go wrong, resulting in an injury, fatality, or property loss. A risk assessment advances the volume of useful information by considering the degree to which a problem will manifest as a function of frequency and severity.

The degree to which hazard identification has occurred must be audited. Job hazard analyses, risk assessments, or other documentation can be reviewed to determine how well risks have been identified in relation to the scope of work and activity that occurs at a given facility.

HAZARD CONTROL

Hazard control occurs as an outcome of the hazard identification process. For example, a job hazard analysis is used to determine what can go wrong in each step of a task as it is performed. Mitigation can be determined for each hazard that has been identified. A job hazard analysis might reveal that maintenance workers are routinely exposed to falls when maintaining a certain piece of equipment. The hazard is controlled by building steps and a work platform with handrails to protect maintenance workers from falling while working on the equipment.

Hazard control measures can be audited through a review of documents and a facility inspection. Documented job hazard analyses and risk assessments can be reviewed, along with hazard control measures that have been identified. These documents can be reviewed for thoroughness and the level of appropriateness in identifying hazard control measures. A facility inspection will reveal how well hazards are being managed in work processes. This activity will determine how well what was done on paper has translated into the work environment.

EDUCATION AND TRAINING

Education and training are processes for communicating hazard identification and hazard control measures to managers and workers. Training is a well-used process of communicating information through classroom sessions and practical skills exercises. For example, a classroom session on lockout/tagout can include a discussion of hazards associated with working on energized equipment and tools to be used to control hazardous energy. Practical skills exercises might include going out into the work environment and going through the process of locking out various pieces of equipment.

Education can include initial training, but it expands to include ongoing efforts to continually reinforce critical information. Information can be included in newsletters, pre-shift meetings, blogs, texts, signs, and posters. There are a large number of educational platforms from which to choose. A review of the organization and work environment can yield ways in which ongoing education will be most effective.

Education and training can be audited through reviewing documents, observing worker behavior, and interviewing workers. A review of documents can be used to assess the occurrence of training and ways in which ongoing education is used in the facility. Observing workers can be used to determine how well they are carrying out tasks, considering procedures presented in training sessions. Interviews can be used to determine how well workers have retained information and are aware of how work should be performed.

MANAGEMENT SYSTEM ASSESSMENT

The management system should be assessed on a routine basis, which is the focus of much of this text. Assessment provides the opportunity to evaluate how well management system components have been developed, implemented, and maintained. As previously mentioned, inspections and compliance audits were early tools used to assess performance. These tools are largely focused on regulatory compliance. The implementation of robust management systems has been identified as a process to protect workers and property more fully beyond the limitations of basic regulatory compliance.

Management systems can be assessed through the processes of documentation review, facility inspection, and worker interviews. A review of documentation will address regulatory compliance issues in the form of written programs and documented training, as well as the use of supporting documents such as confined space entry permits. A facility inspection will address how well written programs and policies are carried out in practice. Employee interviews will reveal how well information has been received and applied by workers.

Perception surveys can be used as an additional tool to assess the perceptions of workers and management in relation to how well the management system is implemented and functioning. Perception surveys are valuable in that they might reveal information that can have a significant impact on performance improvement. For example, conducting a review of documents, a facility inspection, and employee interviews might yield a high score, which indicates the management system is functioning at a high level. However, a perception survey might reveal that employees do not feel management authentically cares about their safety, while the same perception of management reveals that they believe they do care about the safety of their workers. This gap in perception can drive an analysis of why workers feel the way they do and what things management can do to communicate that they care for the safety of workers.

CASE STUDIES

FIRE

Susan was recently hired by the city fire department and is excited about achieving her lifelong dream of becoming a firefighter. She feels competent to engage in all aspects of emergency response operations. After responding to calls during her first two months, she has concerns about scene safety when responding to car accidents on the highway that runs through the city. She has mentioned her concern but has been met with the response that operations have occurred as they have for years and no one has been injured.

SAFETY

Cathy has been working to create a more robust assessment of her facility's safety management system due to her recent promotion to Safety Manager. A recent random OSHA inspection revealed no issues, but she believes more can be done to protect workers and company property.

SECURITY

Steve has managed security at the facility for five years. A contract security company provides security officers who staff strategic posts throughout the property to maintain access control. While walking through the facility, he came across a contractor with whom he was not familiar and stopped to speak with them. He found they were not on the list of approved contractors but were granted access to the facility.

- What issues might exist in the management system?
- What component(s) of the management system can be used to improve organizational performance?

EXERCISES

For the following questions, identify a single facility environment in which you would like to situate your responses and answer each question accordingly. Answer each question in the context of a fire, safety, or security audit.

1. How can leadership be assessed within the context of a management system?
2. How can worker engagement be assessed within the context of a management system?
3. How can hazard identification be assessed within the context of a management system?
4. How can hazard control be assessed within the context of a management system?
5. How can education and training be assessed within the context of a management system?
6. Why is it important to assess the performance of a management system?
7. How can perception surveys be used to assess a management system?

Part II

Defining a Management System

5 Management System Standards

Management system standards exist in the field of occupational safety. These are considered consensus standards, which means professionals in the field have identified and voted on what should be included in standards as a consensus decision. Though these standards are focused on the field of occupational safety, their content is transferable to the disciplines of fire and security. Access to the standards must be purchased through the organization that publishes them.

Much of this text is focused on the development of a management system audit. An exploration of management system standards will assist in gaining an understanding of what comprises a management system. Defining an applicable scope of activity within the management system will be necessary to create an audit program and audit document to assess the performance of the system most effectively. The management system standards that exist in the field of occupational safety include:

- ANSI/ASSP Z10
- ISO 45001
- OHSAS 18001

Exploring these standards will provide the opportunity to understand the similarities between the proposed systems and the activities that should occur when implementing a chosen standard. In addition to these standards, OSHA Recommendations for Safety and Health Programs will be explored. This information was previously referred to as OSHA's Injury and Illness Prevention Program (I2P2). Though OSHA's material is not in the form of a standard, it provides a similar body of knowledge as that of the management system standards. OSHA's material is free to the public and can be accessed at https://www.osha.gov/safety-management.

ANSI/ASSP Z10

The American National Standards Institute (ANSI) and the American Society of Safety Professionals (ASSP) jointly manage ANSI/ASSP Z10 – Occupational Safety and Health Management Systems. As such, ANSI Z10 is a US-based standard, with domestic organizations targeted for its use and application. The primary elements of the management system include:

- Management Leadership and Employee Participation
- Planning

DOI: 10.1201/9781003371465-7

- Implementation and Operation
- Evaluation and Corrective Action
- Management Review

Management is seen as responsible for initiating and maintaining all elements of the management system. They are responsible for the primary areas of establishing the management system, creating policy, and defining responsibility and authority for managing the system through supplying appropriate personnel, budgeting, and integrating the management system into other management systems used within the organization.

Employee participation is presented as critical to ensuring the management system is not simply a top-down exercise when workers are told what to do. Employee participation is stated as needing to occur in at least the phases of planning, implementation and operation, and evaluation and corrective action. Engagement in these phases will ensure employee input on how the management system will be designed, implemented, maintained, and reviewed for effectiveness, with process improvements implemented where needed.

Planning involves a review and assessment of all issues that impact the management system, such as hazards present in the work environment, resources available to support the management system, and areas of compliance. Once all issues have been identified, they can be prioritized to address the most critical issues first. Objectives can then be set to methodically work through addressing the issues in a manner that results in management system improvement. The organization can implement its plan and monitor the success of each area of opportunity that has been identified for improvement.

Implementation and operation involve a broad scope of activities that must be carried out in the daily application of the management system. This includes activities such as conducting risk assessments, utilizing appropriate controls to eliminate or reduce identified risks, and conducting worker training. These and a number of other activities that will be discussed in greater detail later in this text must be carried out for the management system to accomplish its intended aim of reducing fatalities, injuries, and property damage.

Evaluation and corrective action provide the opportunity to analyze the effectiveness of the management system and make corrections when needed. This can be accomplished by implementing and reviewing lagging measures, leading measures, trends identified through evaluating incident investigations, and auditing. A complex system including multiple levels of evaluation will provide opportunities to analyze the management system from multiple perspectives, leading to agile and immediate adjustments as well as long-term systemic adjustments.

Management review refers to upper management reviewing the performance of the management system and facilitating change when needed. Such a review should be performed on at least an annual basis. This review is apart from an audit that would be conducted by a discipline professional, such as a Safety Manager. The Safety Manager can present audit results and other management system measurements and activities to upper management for review and consideration.

ISO 45001

The International Standards Organization (ISO) manages its 45001 standard, titled Occupational Safety and Health Management Systems. The creation and publication of ISO 45001 in March 2018 involved input from 75 countries and was immediately adopted by 63 countries, including the United States. Facility certification to this global standard serves as an international benchmark for safety and health management excellence. The primary elements of the management system include:

- Context of the Organization
- Leadership and Worker Participation
- Planning
- Support
- Operation
- Performance Evaluation
- Improvement

Most of these sections have similarities to the requirements of ANSI/ASSP Z10. The context of the organization is a unique initial point to consider. The organization needs to be fully understood as the context in which the management system will be situated.

Leadership and employee participation are similar to ANSI/ASSP Z10. Leaders must be engaged to the extent of providing resources and incorporating the management system into the overall operation of the facility. Workers should be provided opportunities to participate in the management system as appropriate within the context of the organization.

Planning is similar to ANSI/ASSP Z10. The facility must engage in activities such as evaluating risk, implementing hazard controls, and addressing regulatory compliance.

Support is a unique section but covers similar content in ANSI/ASSP Z10. The intent of this section is to consider critical issues that must be addressed to implement the management system. Issues such as document creation and control, communications, and worker awareness are addressed.

The remaining sections of operation, performance evaluation, and improvement are similar to ANSI/ASSP Z10. Operation addresses the daily and ongoing activities that must occur to implement the management system. Performance evaluation addresses the need to assess the performance of the management system from multiple perspectives. Improvement addresses the need to respond to issues identified in various evaluation processes.

OHSAS 18001

Occupational Health and Safety Assessment Series (OHSAS) 18001, published in 1999, is mentioned in this text only as a matter of historical perspective. OHSAS 18001 was a prior global standard titled Occupational Safety and Health Management

Systems and was managed by the British Standards Institution (BSI). An announcement was made during the creation process of ISO 45001 that OHSAS 18001 would be terminated over a 3-year period of the inaction of ISO 45001 in favor of the new ISO 45001 global standard. Similar to ANSI/ASSP Z10 and ISO 45001, the primary elements of the management system included:

- General Requirements
- OH&S Policy
- Planning
- Implementation and Operation
- Checking and Corrective Action
- Management Review

OHSAS ceased to be a global standard in 2021.

OSHA RECOMMENDATIONS FOR SAFETY AND HEALTH PROGRAMS

OSHA has encouraged organizations to develop a safety and health management system for many years. Regulatory compliance as a sole organizational effort results in limitations to performance. This realization brought about the need to address management system components beyond compliance activities. Early efforts by OSHA were branded as an Injury and Illness Prevention Program (I2P2).

An attempt was made to create a safety and health management system standard that would legally require employers to comply with I2P2 components. This effort was not met with support by industry due to not wanting to be required to comply with certain regulations as well as being mandated as to how such activity should be accomplished. Considering the failed management system regulation, OSHA made a secondary attempt by including compliance with their management system components as a line item in citation settlement agreements resulting from compliance inspections. Safety professionals were soon alerted to this effort and were encouraged to strike that line item from the settlement agreement. Failure to do so would legally obligate the cited employer to comply with OSHA's management system components.

OSHA's management system material is now branded as Recommended Practices for Safety and Health Programs. Though presented in a different format, the content is similar to ANSI/ASSP Z10 and ISO 45001. The primary elements of the management system include:

- Management Leadership
- Worker Participation
- Hazard Identification and Assessment
- Hazard Prevention and Control
- Education and Training
- Program Evaluation and Improvement

- Coordination and Communication for Host Employers, Contractors, and Staffing Agencies

Management leadership, worker participation, hazard identification and assessment, hazard prevention and control, education and training, and program evaluation and improvement were components of OSHA's original I2P2 model. Coordination and communication for host employers, contractors, and staffing agencies is an added seventh component of the current model.

TRANSLATION TO OTHER DISCIPLINES

ANSI/ASSP Z10, ISO 45001, and OSHA's Recommendations for Safety and Health Programs are designed to address occupational safety and health management systems. However, the concepts presented in these standards and guidance can be translated and applied in other disciplines. These standards and guidance can be viewed generally, apart from their specific application to occupational safety. For example, the following can occur:

- Leaders can engage and provide support for security, as well as high-ranking officers being engaged and providing support in the fire service.
- Workers can become engaged in security improvements, as well as firefighters having a voice in management system improvements.
- Security risks must be identified, as well as risks in the fire station and in emergency response.
- Security risk control measures can be identified, as well as those in fire stations and in emergency response.
- Education and training need to occur in the realm of security as well as relevant areas in the fire service.
- Management system evaluation and improvement can occur in the field of security and in the fire service.

Principles in the standards and guidance can be directly translated into the field of security. The fire service holds the opportunity to apply the standards and guidance as they exist as well as translate them to the topic of security. Security can present a unique challenge in the fire service and warrant investigations due to the traditional role of the fire department being open and accessible to members of the community. Such access can present security risks that should be evaluated and controlled in a security management system.

The chapters that follow will utilize the OSHA Recommendations for Safety and Health Programs model to explore the contents of a management system more fully. This is done for two reasons. First, the OSHA model closely follows the content of ANSI/ASSP Z10 and ISO 45001. Second, OSHA's material is available free of charge on their website at https://www.osha.gov/safety-management. This open access provides a platform for discussion that might be limited by paid access to ANSI/ASSP Z10 and ISO 45001. Third, the content of OSHA's model is transferable to other disciplines.

CASE STUDIES

FIRE

Frank is the training officer of his fire department. He notices that a recent increase in training sessions he has had to conduct is due to injuries experienced by firefighters. Training sessions are needed to equip firefighters to backfill injured firefighters who are off duty to heal from work-related injuries. After inquiring about safety protocols in the department, he notices that there are a number of opportunities for improvement.

SAFETY

Susan is the Safety Manager of a manufacturing facility that produces a unique part that is used in the construction of windmills. The company's business has expanded from domestic sales to international sales in five countries. A potential business in a sixth country is in dialog with her Plant Manager, but there is concern about how well Susan's company is managing workplace safety.

SECURITY

Mary has worked tirelessly as the site Security Manager at a distribution center to improve the security management system. She has struggled with how to articulate the key elements the organization should work on to protect its assets and its people. Upper management tends to rely on the basic issues of security officers staffing posts, maintaining access control, and processing employees and visitors in and out of the facility.

- Which standard or guidance would be the best solution among ANSI/ASSP Z10, ISO 45001 and OSHA's Recommendation for Safety and Health Programs?
- How can the standard or guidance best prepare them to move forward?

EXERCISES

For the following questions, identify a single facility environment in which you would like to situate your responses and answer each question accordingly. Answer each question in the context of a fire, safety, or security audit.

1. How can compliance with the components of ANSI/ASSP Z10 add value to the organization?
2. How can compliance with the components of ISO 45001 add value to the organization?
3. How can compliance with the components of OSHA's Recommendations for Safety and Health Programs add value to the organization?
4. How can the safety-specific content of these standards and guidance be translated in the disciplines of security and fire?

6 Leadership and Worker Participation

Leadership and worker participation deserve a cohesive discussion due to the general goal of engaging both management and workers in the management system. Subject matter experts, such as a Safety Manager or Security Manager, can develop programs and training material, but the participation of management and workers is critical for the comprehensive management system to be effective. Leaders must set the tone for the importance of and provide support for the management system within the operations of the organization. Workers must allow the organization to take advantage of their unique knowledge and perspective to improve the management system.

DEMONSTRATE LEADERSHIP

Leaders do not need to be subject-matter experts in safety, security, or fire. However, they do need to be engaged in the management system and provide all the necessary support for it to function at its greatest potential and become integrated into other management systems in place at the facility. Effort must be made to determine the most effective ways for leaders to demonstrate support for the management system that make sense for the organization, as there is no one-size-fits-all solution. Numerous options are available that can be customized for various individuals in leadership roles.

Plant managers can include information in periodic plant meetings. For example, if plant meetings are routinely held once each quarter, safety performance information or communication of safety initiatives can be included in the meeting. Workers perceive what is covered in such meetings as important, which makes these meetings an avenue for leadership to demonstrate their focus on the management system and its importance to the organization.

Critical areas of focus in the management system can be included in overall organizational goals in a given fiscal year. Doing so will assist in incorporating the management system into the general focus of performance improvement. For example, a security goal of reducing product loss by a certain percentage or amount could be included if significant loss has occurred. Workers and management will begin to see security as a natural component of the organization.

Management system activities can be included in the annual performance evaluations of managers and workers. This process can be used to hold individuals accountable for specific activities they can engage in that should foster organizational performance improvement. Caution should be taken to include only things on performance evaluations that the individual has the sole ability to accomplish. For example, placing the goal of a ten percent reduction in injury rate on

DOI: 10.1201/9781003371465-8

a departmental supervisor's performance evaluation can be problematic. They cannot individually accomplish that goal due to the behavior of each worker in the department affecting the occurrence of injuries. However, managers can do things to influence the reduction of an injury rate, such as ensuring each worker has received the appropriate safety training and including safety in pre-shift meeting discussions. Converting those activities into goals is something managers can independently accomplish.

Managers can engage in routine activities that occur within the management system. Rather than all activities being relegated to the subject matter expert, such as a Security Manager, managers can assist in accomplishing certain tasks. Activities managers can engage in include things such as:

- Assisting with safety, security, and fire inspections of their department
- Participate in incident investigations
- Participate in management system audits
- Include information related to the management system in plant, shift, and pre-shift meetings
- Encourage workers to become engaged in activities like safety committee membership
- Make recommendations for management system improvements
- Ensure all workers within their scope of responsibility have received the appropriate training
- Engage in professional development in topics related to the management system

MANAGEMENT COMMITMENT

Oregon OSHA examined the issue of management commitment to safety and generated a list of recommendations that can be used to engage organizational leaders in the management system. Recommendations include:

- Upper management can publish a safety policy that clearly articulates the commitment of the organization to worker protection. This policy can be communicated during new hire orientation as well as posted throughout the facility for ongoing reference and awareness.
- Managers at all levels of the organization can make sure workers follow safety procedures. A positive culture can be developed that incorporates open communication regarding the need to follow safety procedures. Such communication can reveal previously unidentified barriers to workers performing tasks safely. For example, a worker might perform a task unsafely through no fault of their own. A conversation might reveal a lack of training or undue pressure placed on the worker by the organization to cut corners in the production process.
- Upper management can give workers the authority needed to carry out work safely. Workers can be provided avenues to report hazards they encounter. They can also be provided with the ability to shut down work if a significant

safety hazard has been identified. Safety committee members can be provided with the time and financial resources to engage in incident prevention activities.

- Upper management can allocate financial resources to support the implementation of the management system. This will require attention to resources that involve personnel, capital expenditures, and minor purchases. Salary and benefits might be needed to hire a safety manager and additional safety staff. A substantial amount of money might be needed to retrofit a work area with steps, working platforms, and handrails where fall hazards have existed. A smaller amount of money might be needed to purchase safety glasses for workers in certain areas of the facility.
- Upper management will need to budget the time needed to implement the management system. Time must be taken to conduct audits and other management system measurement activities. Workers will need to be provided time to be away from work for safety training and participation in the management system, such as attending monthly safety committee meetings.
- Management at all levels in the organization needs to act on worker recommendations for performance improvement. Though no action might be able to be taken immediately, leaders can maintain open lines of communication with workers on why certain things might not be able to be addressed while also keeping them updated on progress made toward recommendations that can be addressed.
- Manager must make sure all workers under their scope of responsibility have received the appropriate safety training to work safely.

A university president once said, "Everything rises and falls on leadership." At first glance, that statement might appear overly simplistic due to the complex nature of organizational operations. Yet organizational leaders hold great power in influencing organizational culture and setting the tone for how work is accomplished. Leaders can engage in strategic activities to support the integration of safety, security, and fire management systems.

BENEFITS OF WORKER ENGAGEMENT

Similar to leaders needing to be engaged in the management system, so must workers. Like leaders, the avenues of engagement and degree to which workers are involved should be viewed as agile based on the organization and desires of workers. One facility might contain a worker population that tends to desire to be highly involved, while another facility might contain a worker population that tends to have little desire to be engaged. A review of the facility should first include an understanding of the degree to which workers want to become involved. The degree might also evolve over time. For example, a plant might contain a population of workers who tend not to want to be engaged in the management system. A few workers might be interested in participating in processes. Over time, other workers might see the value of participating in the management system, which can change the dynamics that initially existed. Each facility should be assessed independently, with flexibility for worker engagement to change over time based on their desires.

There are a number of beneficial reasons to engage workers in the management system. Some of these benefits include:

- Worker engagement is a fundamental principle of leadership. Worker engagement establishes a partnership in operations as opposed to top-down processes where workers simply perform task, as they are told. Worker engagement elevates their role in the organization by giving them the opportunity to communicate with leaders about performance improvement opportunities.
- Worker engagement instills ownership in business processes. Workers can feel they are a greater part of the business and be more vested in management system initiatives when they participate.
- The organization can experience substantial performance improvement by capitalizing on the knowledge and experience of its workers. Workers hold special knowledge of how work is performed and the factors that influence worker performance. The feedback they provide can enlighten leadership on issues that were previously unknown.
- Research has shown that worker engagement is a safety success factor. Workers can provide subject-matter experts with a wealth of information that can be used to reduce the occurrence of incidents.

INFORMAL AND ANONYMOUS OPPORTUNITIES

Workers can be engaged through the basic process of reporting hazards. They are positioned to be exposed to or observe hazards that exist throughout a facility. The temperament of workers and the culture of the organization can impact how comfortable workers feel coming forward with a report of a hazard. A solution to this challenge is to consider the different avenues that are available to foster the hazard reporting process. Options can include the following:

- Workers can report issues or make recommendations directly to management. Workers can approach their supervisor and have an open dialog about issues that have been observed or experienced. A very positive organizational culture must be in place for this behavior to occur.
- When workers do not feel comfortable reporting issues to management, they can report issues or make recommendations to a senior peer or their peers on a safety committee. They may not feel comfortable approaching management, but they do feel free to engage with their peers.
- A suggestion box or other platform can be provided for workers to report issues or make recommendations anonymously. Such a desire among workers might not necessarily reflect a poor organizational culture. Workers might simply be shy and prefer a completely anonymous format to engage in the process.

SAFETY COMMITTEES

Similar to the recommendation for leadership engagement, Oregon OSHA explored the issue of worker participation and made recommendations for how workers can become engaged through safety committees. Their recommendations include:

- Define membership based on the groups represented in the facility. A pro-rated distribution of committee members can be established based on the volume of workers in various areas. Consideration can also be given to management and contractors who are onsite during most operational hours, such as contract security and food service. For example:
  - 8 workers prorated among worker populations in various departments
  - 1 representative from the contract security company
  - 1 representative from the food service company contracted to staff the cafeteria
  - 1 member of management

The size of the facility might dictate the need to have one committee per shift to expand representation among employee groups and provide more opportunities for workers to participate while keeping committee size to a reasonable number of participants.

- Requirements should be set for membership on the committee. Requirements can include things such as:
  - Strong record of attendance at work
  - No corrective action within the past 12 months
  - At least 1 year of employment with the company
  - Acceptable work performance

Setting standards for committee membership will ensure the committee has a positive image by including workers and managers who are deemed to be strong performers in the organization. It will also help to avoid attracting individuals who might engage in a committee only for the personal benefits they might experience from committee membership, such as committee t-shirts or meeting refreshments.

- Responsibilities should be delineated for each member of the committee. Each member should be fully aware of the commitment that must be made to the committee through specific activities in which they must engage.
- The committee should set goals for what is to be achieved in a given fiscal year. Incidents can be reviewed to determine potential opportunities to set goals for performance improvement in certain areas of the operation.
- The committee should meet on a regular basis. Committees typically meet monthly. Management and workers can communicate to establish the appropriate basis for the committee to meet.

- The committee can collect safety suggestions from workers. The avenue for accomplishing this can be defined by the committee and communicated to all workers and managers so recommendations can be made in a timely manner.
- The committee should act on suggestions received from workers and managers. The level of action will vary based on the context of the recommendation. For example, a recommendation might be made that will require a capital investment. The recommendation might involve work activity that is currently managed in a safe manner, but the recommendation increases the protection of workers. Immediate action might not be able to be taken and the committee can communicate with workers and managers indicating that the recommendation has been received, evaluated, and cannot be acted on soon, but it is being held in a queue for future process improvements when funding is available. The committee can similarly communicate to workers and managers when recommendations are acted on and process improvements have been achieved. Acting on suggestions and maintaining open communication with workers and managers can assist in establishing the integrity of the system.
- Someone on the committee should be assigned to take meeting minutes. The process of taking meeting minutes will help to establish a record of things that have occurred within the committee and future actions that should be taken.

INCIDENT INVESTIGATIONS

Incident investigations serve as an opportunity for workers to participate in the management system. They can engage in a variety of inspections to include injury, property damage, and security-related incidents. Workers can engage in the following ways:

- Accompany a supervisor or Safety Manager during the investigation process
- Analyze the scene and provide a unique perspective on how work is performed
- Provide input on what might have caused the incident given their experience performing actual or similar tasks
- Recommend corrective action that can be implemented to prevent the future occurrence of similar incidents
- Encourage workers to exhibit safe work behavior based on lessons learned from incident investigations

AUDITS AND INSPECTIONS

Experienced workers can become engaged by participating in audits and inspections. Experienced workers can engage in conducting them, while less experienced workers can be engaged by being the subject of audits and inspections. Workers can engage at varying levels, including the following:

- Less experienced and experienced workers can accompany a supervisor or Safety Manager while an audit or inspection is being conducted. The supervisor or Safety Manager can leverage the event as a training opportunity by offering insight on how to provide a safe work environment most effectively. Workers can provide input on operational issues that impact the degree to which work can be performed safely.
- Workers can engage with an auditor or inspector while the audit or inspection is being conducted. They can escort the auditor or inspector through their department and respond to questions that might be posed regarding safe work operations.
- Experienced workers can be trained to conduct audits and inspections autonomously. They can receive training on risks to consider and specific items to be evaluated while conducting the audit or inspection. They can be trained on the content of the audit or inspection document and how to properly enter information and process the completed document.

CASE STUDIES

FIRE

Jenny was hired by the city to address the high rate of injuries among city employees. The fire department rose to the top of her list of problem areas due to the volume of injuries and workers' compensation costs. Further investigation yielded a hostile relationship between the firefighters' union and city government.

SAFETY

Ahmed is the Safety Manager for a motor manufacturing company. The organizational culture is saturated with a top-down mentality where workers are expected to carry out the mandates of management. Injury rates have been shown to have consistently increased over the past fiscal years. Workers respond well to Ahmed's efforts in communicating with them as to how to improve things, but workers are apprehensive about management.

SECURITY

Chloe is the Security Manager of a distribution center and has identified an increase in product loss after reviewing recent inventory control reports. She pursued areas of investigation and found the loss was due to theft. The facility has gone through a number of years without cost of living pay raises, which caused her to consider if theft is occurring in retaliation as a way to offset what workers believe they are due, as there is a positive correlation between years without a pay increase and an increase in product loss.

- How can leadership foster management system improvement?
- How can worker engagement foster management system improvement?

EXERCISES

For the following questions, identify a single facility environment in which you would like to situate your responses and answer each question accordingly. Answer each question in the context of a fire, safety, or security audit.

1. How can leadership impact the performance of the management system?
2. What things can management do to demonstrate their commitment to the management system and its improvement?
3. How can worker engagement impact the performance of the management system?
4. What can workers do to engage and improve the performance of the management system?

7 Hazard Identification and Control

Hazard identification is a component of a management system that provides for the ability to evaluate the workplace to determine where hazards to people and property exist. Hazard control is the component of the management system that provides for the determination of what method or methods can be employed to prevent an incident from occurring that is related to each hazard. Hazards to people and property exist in the realms of safety, security, and fire. Flexibility can be used to adapt solutions to address hazards that are unique to each organization.

DEFINITION OF A HAZARD

OSHA defines a hazard as "…the potential for harm. In practical terms, a hazard often is associated with a condition or activity that, if left uncontrolled, can result in an injury or illness." Hazards can vary in nature within the scopes of safety, security, and fire. Examples of safety hazards include things such as:

- A machine with moving parts that are not guarded and present the ability for a worker to have a body part pinched or removed
- A missing section of handrail that can cause a worker to fall to a lower level
- Sanding a wood or metal surface that results in particulates being emitted into the air
- A confined space that contains an air contaminant
- A work area that contains machinery that operates at an exceptionally loud level
- A cleaning material that contains toxic ingredients
- Working with heavy material that could fall on a worker's foot
- Repetitively using the same muscle group throughout a shift

Examples of security hazards include things such as:

- An unmonitored building access point
- Uncontrolled departure of contractors from a facility
- Entering a facility without the contents of bags being examined
- Unauthorized individuals having access to a facility
- Ability for individuals to bring a weapon into a facility
- Unapproved individuals riding in a vehicle onto company property
- Visitors who are unaccompanied by a representative of the company
- Gap in perimeter fencing

DOI: 10.1201/9781003371465-9

Examples of fire hazards include things such as:

- Improperly stored chemicals
- Excessive presence of flammable material that is beyond the capability of the fire suppression system
- Lack of accountability system on fire response scenes
- Poorly maintained electrical systems
- Overloaded electrical systems
- Unprotected flammable material while cutting and welding is occurring
- Excessive accumulation of flammable dust
- Material storage that blocks a fire suppression system sprinkler head

Hazards may be shared among different workplaces, while some hazards might be unique to a given workplace. For example, security hazards related to access control might be shared among a broad spectrum of facilities. However, the accumulation of flammable dust will be unique to certain types of facilities, such as grain elevators used to store grain purchased from farmers. The movement of the grain throughout the facility generates grain dust that can pose a risk of fire and explosion.

JOB HAZARD ANALYSIS

A job hazard analysis (JHA) is a tool used to examine work as it is performed to identify risks that are present. Though a JHA has historically been used in the field of safety, it can be translated for use in security and fire. A job hazard analysis is a technique that examines each step of a task as it is performed to determine the risks present in each step of the process. This intentional and systematic review can be performed by one or more individuals. Multiple individuals performing a JHA are preferred due to the potential to observe a task from numerous perspectives. For example, a safety manager might view the task clearly as performed, whereas a worker might view the task more dynamically by considering things that might not be observed during the JHA.

A job hazard analysis can be conducted by using a form to record information as observed throughout the process. The first objective is to break down the task into its primary steps. The number of steps will vary depending on the task being performed. The second objective will be to observe each step of the task and identify risks to people and property that are present in each step. Observers can focus on what can go wrong and the catalysts for its occurrence.

HAZARD CONTROL

Once each step of a task has been observed and hazards have been identified for each step, control measures for each hazard will need to be identified. Control measures can be selected by navigating the hierarchy of controls. The hierarchy of controls provides a rank order of control methods that range from most preferred to least preferred. Methods available in the hierarchy of controls include:

- *Elimination*: Eliminating the hazard is the most preferred control due to the hazard becoming non-existent in the workplace. For example, a JHA or risk assessment might reveal the exposure of maintenance workers to the hazard of electrocution when working on facility electrical systems. Electrical work is conducted on a non-routine basis. This hazard can be eliminated among maintenance workers by contracting the work to a local electrical company. The result of this control is that it completely removes exposure to the hazard from maintenance workers who engage in the work on a non-routine basis while allocating the work to a third party that engages in the work routinely and will be better positioned to perform it safely.
- *Engineering*: Engineering controls are a form of hazard elimination in that the hazard is eliminated by designing the work environment so that it no longer exists. For example, a JHA or risk assessment might reveal that maintenance workers are exposed to a fall hazard while performing preventive maintenance on a certain motor that powers a conveyor belt. An engineering solution is to design and install a catwalk system that includes steps, a work platform, and handrails. Maintenance workers can then easily access the motor and perform maintenance from the safety of a protected working surface. "Prevention through design" is a process that can be used in the design phase of new building construction or expansion. Safety, security, and fire professionals can be included in the design process to assist in identifying potential hazards that can be minimized or eliminated in the design process.
- *Administrative Controls*: Administrative controls involve work procedure alterations that can control an identified hazard. For example, a JHA or risk assessment might reveal that production workers are exposed to the ergonomic hazard of repetitive motion by using the same body part to perform work throughout a shift. The ergonomic hazard can be controlled by rotating workers to other tasks throughout a shift that allow them to utilize multiple muscle groups.
- *Personal Protective Equipment (PPE)*: PPE is the least desirable control but might often be the most frequently selected due to its low direct cost. For example, purchasing fall protection harnesses and lanyards might have a lower direct cost than engineering stairs, working surfaces, and handrails. PPE includes items such as the following:
  - Hard hats
  - Safety glasses
  - Goggles
  - Face shields
  - Respiratory protection
  - Hearing protection
  - Safety-toe shoes
  - Gloves

Selection of PPE should also consider worker behavior from the perspective of selecting PPE that is most likely to be worn. For example, foam ear plugs might be a low-cost option to address the hazard of high noise, but earmuffs might prove to be more comfortable for workers to wear and maintain.

RISK ASSESSMENT

Risk assessments elevate hazard assessment and response to include a measurement of risk as a function of frequency and severity. Though JHAs remain a common component of hazard identification and response, organizations are evolving their processes in favor of risk assessments due to the additional information provided by the activity. A risk assessment can result in tasks being placed in a matrix that measures the degree to which workers are exposed to an incident. A fundamental matrix can be created that plots tasks into one of four categories:

- *Low*: low frequency of performing the work and low severity of injury or property loss if an incident occurs.
- *Moderate*: low frequency of performing the work and high severity of injury or property loss if an incident occurs.
- *Moderate*: high frequency of performing the work and low severity of injury or property loss if an incident occurs.
- *High*: high frequency of performing the work and high severity of injury or property loss if an incident occurs.

Plotting tasks on a matrix with these sections provides a clear list of priorities to respond to by implementing control measures. High-risk tasks can be addressed first, followed by moderate-risk tasks, and then low-risk tasks. Detailed information regarding conducting risk assessments at the introductory, intermediate, and advanced levels can be found at https://www.assp.org/advocacy/risk-assessment-committee. The American Society of Safety Professionals (ASSP) has posted open access material to increase the skill set of professionals who wish to integrate risk assessment processes into their organizations.

BENEFITS OF A CONDUCTING JHAS AND RISK ASSESSMENTS

There are numerous benefits of conducting JHAs and risk assessments. JHAs and risk assessments can influence reduced losses to the company and increased productivity. Some of these benefits include the following:

- Fewer injuries should be experienced through identifying risks to injury and controlling worker exposures.
- Workers' compensation costs should be reduced by lowering the frequency and severity of injuries.
- Increased productivity can be achieved through maintaining a safe environment for high performing, experienced workers.
- Can be used as a tool to construct task procedures that maximize safety and productivity.
- Can be used as a training tool to orient workers to a new task.

- Maintain continuity of plant operations through the prevention of fires.
- Provide for the safety and security of workers and managers through effective access control procedures.
- Property loss should be reduced through reduced theft.

WHERE TO START

Conducting JHAs and risk assessments can seem to be an overwhelming undertaking when considering the volume of tasks performed at a given facility. Conducting a JHA or risk assessment and responding to issues can be a resource-intensive process, to include the time required to conduct them thoroughly and respond to issues that are identified. A process can be established to systematically work through each task at a facility in a prioritized fashion.

A first reactive step is to review all incident reports within the past year. A spreadsheet or database can be created to record critical pieces of information. Columns can be created to record:

- Date of incident occurrence
- Department in which the incident occurred
- Location of the incident
- Shift on which the incident occurred
- Brief description of the incident
- Causes of the incident
- Dollar amount of loss
- Production impact on the facility

Analyzing these data points will yield valuable information. Trends can be identified as well as where high-severity and high-frequency losses are occurring. Tasks that rise to the top of the list as having the greatest negative impact on the facility can become the priority for conducting a JHA, or risk assessment.

A second step can be to speak with workers to determine where they believe vulnerabilities exist. Workers engage in tasks that cover every aspect of the operation, which makes them uniquely positioned to provide feedback on hazards that exist in the facility. They can provide insight on issues that potentially exist in safety, security, and fire. Their feedback can serve as a second group of tasks on which to perform a JHA or risk assessment.

A third step can be to methodically conduct a JHA, or risk assessment, on all remaining tasks in the facility. The remaining tasks can be prioritized based on an initial review of the perceived risk of loss. A JHA, or risk analysis, can be conducted in order from higher perceived risk to lower perceived risk.

THE AUDIT PROCESS

JHAs and risk assessments should be audited on a routine basis to ensure the information included remains accurate given the dynamics of the work environment. The cycle of review might be quarterly or annually based on the importance of the JHA or risk assessment in preventing loss to people and property. Reasons to audit JHAs and risk assessments on a routine basis include:

- *Jobs Change:* The way in which work is done can change based on evolution in production efficiency or quality improvement. These changes could impact hazards that workers might encounter.
- *Products Change.* Procurement sources might source products through new vendors. For example, single-source supplying might be utilized to leverage purchasing power in an organization, and this decision might result in different chemical products being introduced in the work environment. These changes could result in new hazards that must be evaluated.
- *Equipment Change:* Work processes might be improved through the introduction of new equipment or facility expansions. New equipment and worker interaction with the equipment could introduce new hazards that need to be analyzed and controlled.
- *People Change.* Turnover, generational traits, and an aging workforce are variables to consider in relation to how changes in people might impact JHAs and risk assessments. Frequent turnover in certain positions might increase the likelihood of the occurrence of an incident. Work can be done to examine why turnover is occurring with efforts made to stabilize workers performing the task. Generational traits can impact the degree to which workers perceive and respond to hazards. The risk of injury might increase as workers age. Safety, security, and fire professionals can engage with human resources professionals to examine the legal and practical implications of the interaction of people and hazards.

Including JHAs and risk assessment in an audit strategy can help ensure emerging hazards are identified and responded to in a timely fashion.

REPORTING HAZARDS

A system should be in place for workers and managers to report the presence of hazards beyond JHA and risk assessment processes. A reporting system will allow for the identification of and response to hazards on an ongoing and agile basis. Reporting hazards should be included as a fundamental management system requirement. The importance of reporting hazards that workers identify should be clearly communicated in the new-hire orientation phase as well as being routinely revisited. Workers should be encouraged to report hazards without fear of repercussion.

Workers should receive training on hazard identification. It cannot be assumed that workers will naturally be able to identify things that can result in injury to people and property loss. For example, some industry sectors might stereotypically foster work to be performed in an unsafe manner, such as working from heights without the use of fall protection. Workers who mature in such a work environment might not perceive anything wrong with such behavior. Education and training on hazard identification and control must occur to make everyone aware of work expectations.

Hazard reporting can be carried out through mechanisms that are deemed appropriate for the organization. Options for hazard reporting can include the following:

- Hazards can be reported to a member of management. The member of management can immediately forward the information to the individual(s) who are positioned to address the hazard.
- In the event a worker might not feel comfortable approaching a member of management, a report can be made to the appropriate committee, such as a safety committee. The committee can then forward the report to management to be addressed.
- If a committee is not present or a worker does not feel comfortable making a report to a committee, a report can be made to a co-worker who is comfortable forwarding the information to a member of management or a committee.
- An anonymous mechanism can be used to collect hazard reports, such as a physical or virtual box in which workers can deposit a basic form that delineates the nature of the hazard and where it exists.

Multiple avenues of reporting can help ensure ways are present that will make a worker feel comfortable reporting a hazard. This will be particularly important if hazard reporting is mandated within the management system. If workers are required to report hazards, multiple avenues should be in place to facilitate hazard reporting.

Managers should be delineated with specific responsibilities in the hazard reporting process within the management system. Similar to workers, managers should receive training in hazard identification. Regardless of their prior work history, it cannot be assumed that they have the skill set to identify hazards in the workplace. Managers should then direct the hazard reporting process. They should encourage and support workers in reporting hazards as well as making reports of hazards as observed. Managers should act on hazard reports by forwarding information to the appropriate personnel in a timely manner. This is of critical importance not only in protecting workers but also in managing general risk to the organization. Once a manager is aware of a hazard, timely action must be taken to control the hazard prior to the occurrence of an incident. Upper management is responsible for issues on a strategic level. They will need to engage in activities such as:

- Provide financial resources to address hazards that have been identified in the workplace
- Provide equipment necessary to control hazards at the most effective level, from engineering hazards out of the system to the use of PPE
- Hire subject matter experts, such as safety professionals and security professionals, who can direct the management system and coordinate hazard identification and control

A documentation process should be created and implemented to manage the flow of information related to hazard reporting and response. The documentation process provides a written record that can be tracked from the initial report to the final resolution. A documentation process can include the following:

- Submission process where the hazard is documented as to its type and location in the facility. This information can be documented in a formal report to a member of management or a committee as well as an anonymous submission online or in paper form.
- If hazards are reported to a committee, meeting minutes can be used to track the initial report and the action that occurred.
- Documentation can be used to ensure each hazard has been addressed or placed in a queue for a later response if the severity of exposure is low.
- Documentation can be periodically analyzed to determine trends that might exist. This information can be used to focus attention on certain work areas or tasks that appear to recur among reports.

CASE STUDIES

FIRE

Elly is a firefighter and is concerned about scene safety when responding to automobile accidents on the highway. Reports of first responder injuries and fatalities from across the country in such environments are causing her concern. She has also experienced a number of close calls where motorists who were not paying attention almost struck responding personnel and equipment, as well as motorists who drove near the response scene at a high rate of speed.

SAFETY

John is a new hire maintenance manager at a manufacturing facility. He is an experienced maintenance technician, having worked in industry for different employers for 20 years, and this is a step up in his career. A new experience for him is the need to engage in and manage electrical work beyond what he has typically done for prior employers, where he only engaged in basic activities such as replacing fixtures or adding outlets. He is concerned with his ability to engage in and manage electrical work at the level his new employer would like him to perform.

SECURITY

Judy is a security officer at a hospital. There has been a rise in acts of violence among healthcare workers in the emergency room, and she would like to do something about it. The hospital administration has been supportive of prior security improvements, and her supervisor has a proven track record of improving the security management system.

- How can JHAs or risk assessments be used to address the situation?
- What reporting process would be most appropriate to implement?

EXERCISES

For the following questions, identify a single facility environment in which you would like to situate your responses and answer each question accordingly. Answer each question in the context of a fire, safety, or security audit.

1. What types of hazards typically exist?
2. How can the process of a JHA improve the protection of people and property?
3. What control methods in the hierarchy of controls can be used to address the hazards?
4. How can the implementation of risk assessments elevate the value of the hazard identification and control process?
5. What are the benefits to the organization of conducting JHAs or risk assessments?
6. What process can be used to prioritize and methodically work through all tasks throughout the facility?
7. How should JHAs or risk assessments be routinely audited to ensure updated information is available to address hazards?
8. What reporting process should be in place to facilitate the identification and communication of hazards in the workplace?

8 Education and Training

The education and training component of a management system provides the opportunity to communicate information identified in the hazard identification and hazard control components to workers and managers. Training is typically defined as a forum in which structured information is delivered to equip workers and managers to properly perform their jobs. Though this is also considered a form of education, there are multiple other platforms that can be employed in the management system to foster ongoing education regarding issues associated with the system.

ADULT EDUCATION

Basic concepts of adult education need to be understood for education and training efforts to be effective. An individual might be considered an adult from a legal perspective at the age of 18 due to that being a point in life where independent life activities begin to occur, such as moving away from home, beginning a career, starting college, obtaining an unrestricted driver's license, and joining the military. A second age at which an individual might be considered an adult is the age of 21 due to being able to purchase and consume alcohol. However, the field of adult education defines an adult from a neurological perspective in that the brain has completed development at the approximate age of 25. Principles of adult education are helpful in developing education and training experiences due to the large portion of many workforces being at the age of 25 or older.

Multiple generations of adults present in the workforce can create a challenging dynamic when designing education and training experiences. Learning styles and traits differ among generational groups, yet numerous generations might be present in a given facility. Generations include:

- Baby Boomers (1946–1964)
- Generation X (1965–1979)
- Millennials (1980–1994)
- Generation Z (1995–2012)

Generation Alpha, born 2013–2025, is the coming generation for future research and consideration in relation to the dynamics of adult education. The volume of generations present in a given organization is affected by an aging workforce, with individuals remaining in the workplace longer in current years. Retirement age is lengthening due to numerous economic and social issues, which adds a fourth generation of Baby Boomers to consider in the near future in addition to Generation X, Millennials, and Generation Z.

Learning styles may range from Baby Boomers, who prefer small groups and sharing their experiences, to Generation Z, who prefer independent learning with visual and active components. The challenge is to create education and training

  DOI: 10.1201/9781003371465-10

programs that meet the needs of various generational groups that coexist in the same facility. Defining a path forward will require considering two principles of adult education that will be of particular use.

First, adults prefer to have control over their learning. Adults obviously cannot be provided with complete control over education and training content. For example, a worker who engages in permit-required confined space entry must receive training with content specified by OSHA. However, adults can be provided control over how they receive training. Some workers might prefer face-to-face sessions that involve an instructor-led session with physical exercises. Other workers might prefer to navigate information independently on an electronic device with follow-up from a trainer. Information can be delivered through multiple platforms to meet the learning needs of various generational groups.

Second adults want to know what a training session means to them. They need to know why the training content is important and how it applies to them on the job. This connection must be intentionally designed into the beginning of a training session to immediately engage adults in the learning process. Failure to do so can result in adults not paying attention to or valuing the information that is presented.

Widely used education and training methods clearly violate the principles of adult education. Two of the most common methods are the improper use of PowerPoint presentations as the sole platform for information delivery and the sole use of videos. The use of both can satisfy a legal or company policy obligation to conduct training, but the larger issue is the consideration of whether knowledge was effectively transferred and workers and managers truly learned critical information needed to perform their jobs properly. PowerPoint presentations are often used as a visual outline of the material covered in a training session. Each slide might contain a list of points that address a given concept. The trainer lectures by navigating the content of each slide. The use of videos as a solo platform for training delivery could be viewed as the least preferred method of information delivery. A trainer might simply play a video in front of a class of observing trainees with no interaction, and the video content is used as the sole avenue for information delivery. Neither the solo use of PowerPoint nor the solo use of videos satisfies the needs of adult learners.

A more desirable model might be to consider training as a process rather than an event. Training is often seen as an event in that a session is held in which workers, managers, and workers attend, and they are considered "trained" once the session is complete. A better approach to supporting the implementation of the management system is to consider education and training as an ongoing process that most effectively ensures workers and managers receive and apply information. This can be accomplished by viewing training and education in two phases.

PHASE 1: COMPLIANCE TRAINING

Select the right training environment to reach workers most effectively. The training could include the following:

- A well-lit and organized training room
- An area on the plant floor where practical skills exercises can be arranged

- A web-based platform that allows workers to independently navigate information

The use of PowerPoint can be useful in a well-lit training room if integrated with adult education in mind instead of only information transference. PowerPoint can be used to provide a review of what is required based on the topic of the training session. For example, a safety training session must occur for maintenance workers and managers on lockout/tagout. Power Point can be used to cover specific requirements of the lockout/tagout program in the form of a lecture. But the Power Point presentation can be greatly enhanced using the following:

- Slides can present questions that are periodically posed and to which trainees can respond. This activity allows trainees to reflect on the training session content and how it can be applied in the workplace.
- Slides can present problem-posing scenarios that challenge trainees to think through issues and discuss them with each other or interact with the trainer on variables that affect the presence of the problem and potential solutions. This activity allows trainees to actively engage in the learning process, even though it is occurring in a training room.
- Slides can present a detailed situation that sets the stage for role-playing to walk through a specific situation. Care should be taken to ensure role-playing is a comfortable and beneficial activity for the audience in attendance.

Web-based training tools can be used as a training platform for workers and managers who prefer to navigate initial training content independently. Sessions can be designed internally or identified through training organizations that offer fee-based access to their web-based training sessions. An investigation should be conducted into the delivery of web-based training sessions to ensure they offer a spectrum of learning tools rather than simply serving as an alternative to a video. Web-based training sessions can utilize a spectrum of learning activities that engage the trainee, such as posing periodic questions that engage the trainee in the session. Web-based training platforms can include the following:

- Open access to eTools on OSHA's website
- Online courses designed by training product vendors
- Internally designed interactive online training sessions should include dynamic use of PowerPoint that can be used to incorporate videos and trainee responses to prompted questions in an auto-running format
- Virtual reality training sessions that place trainees in simulated work settings where they must navigate applications of the material presented in the session

Phase II: Skill Development

Ongoing skill development and reinforcement can be used as a second phase that presents training as an ongoing process. An individual development plan (IDP) can be created that addresses the specific development and training needs of each worker and manager. Integrating the use of IDPs can address the issue of various learning traits and needs among a spectrum of generations represented in a workplace. Rather than requiring all workers to be processed through the same training session, a reasonable spectrum of training platform options can be provided to assist in ensuring that learning occurs among a diverse workforce.

Proper training environments can be selected for ongoing training. These can range from well-lit and organized training rooms to dynamic web-based platforms. The spectrum of generations present and their desired learning styles can drive the selection of training environments.

Practical skills exercises can be used to reinforce information that was learned in initial training sessions. Exercises can be designed and delivered in group or individual sessions. Workers and managers can be observed and assessed on their ability to effectively execute the skills needed to perform their jobs properly.

Performance feedback should be provided to trainees. A test might be used to assess comprehension following a training session. A test score is one type of feedback, but verbal and written feedback can be used to provide detailed feedback to trainees. For example, verbal feedback can be immediately provided to workers following a practical skills exercise. Detailed and specific feedback can be provided that allows trainees to clearly understand how they performed well in addition to opportunities for improvement.

Ongoing training sessions can also serve as a leading metric to assess management system performance on an ongoing basis. The assumption is that effective training experiences will predict the success of the management system. For example, conducting effective safety training sessions should predict the prevention of incidents that can cause injury or fatality to workers and managers, as well as the prevention of damage to company property. The volume and quality of training sessions can be used as a leading measure that can impact the performance of the management system.

A challenge is to transition to a different education and training paradigm and become comfortable with everyone not receiving the same training in the context of information delivery. They will receive the same training in terms of content but can receive it in various formats. A multi-generational workforce will necessitate the exploration of training delivery platforms that most effectively meet the learning needs of the trainees. Each organization will need to explore the degree to which training options can feasibly be provided to workers and managers.

REGULATORY TRAINING REQUIREMENTS

Organizations will need to explore regulated training requirements as a matter of corporate responsibility. Federal and state government agencies may dictate worker and management training requirements in certain regulations. For example, OSHA dictates the frequency and content of training under certain regulations. OSHA's powered industrial truck standard is an example of having detailed training requirements in 29 CFR 1910.178(l):

- Training program implementation
- Training program content
- Refresher training and evaluation
- Avoidance of duplicative training (previously received)
- Operator certification

This OSHA standard is very prescriptive regarding training, whereas others might provide more flexibility as to the content and delivery methods.

Training records will need to be maintained to verify that training was conducted. Options for documenting training sessions include the following:

- A training log will typically contain the signature and printed name of attendees present during the session, the date of the session, and the name of the trainer.
- A test can be administered and maintained that includes one or a combination of multiple choice, true/false, and open response questions.
- A certificate of completion can be used that stipulates the topics covered in the training session, the signature and printed name of the trainer, the signature and printed name of the trainee, and the date on which the training session was conducted.

ONGOING AVENUES OF EDUCATION

Ongoing avenues of education can be used apart from a structured training process. A strategy of ongoing education can be used to support and reinforce information received in formal training as well as communicate new information or initiatives associated with the management system. Opportunities to incorporate ongoing education can include the following:

- Refresher training can be conducted on all topics that pertain to management and workers within the scope of the management system. Though regulations might dictate refresher training only on certain topics, work can be done to go beyond regulatory compliance to include refresher training on all relevant topics.
- Pre-shift meetings can be used as a platform to disseminate information on a consistent basis. Material can be scripted and provided to department supervisors to review with workers related to current issues that impact the performance of the management system. A schedule of topics can also be planned and delivered to workers throughout a given fiscal year.
- Incident reviews can be communicated to workers and managers with a focus on lessons learned from incidents. Opportunities for improvement can be solicited from workers and managers as a benefit beyond the act of education.

- Electronic or hard-copy newsletters can be used to present detailed educational information. Information can include things such as case studies of incidents, management system performance metrics, reviews of certain program components, and highlights of best practices that foster the success of the management system.
- Signs and posters can be used to brand the management system throughout the facility and communicate significant pieces of information that can impact work performance. For example, a "top ten list" can be created and branded that addresses the ten most important behaviors needed to ensure the success of the management system.
- Special events can be held that create unique interest in and awareness of the management system. For example, rather than conduct routine refresher training, work can be done to structure an engaging and interesting activity that draws interest among managers and workers in the management system.
- Texts can be sent to workers and managers as a brief reminder of critical components of the management system.
- Website information can be provided that engages workers and managers in management system information that includes text and videos that will be of interest.
- Videos can be created in the form of Ted Talks or TikTok videos that focus on interesting components of the management system. Technology and creativity can be used among managers and workers to deliver engaging content that can be delivered to members of the organization through text, e-mail, social media, or a website.

Numerous opportunities exist for organizations to be creative and inspiring in the delivery of management system educational material. Evolutions in technology have made unique avenues of education accessible to individuals. Interesting content can be created and delivered that can provide information needed by workers and managers to support the ongoing implementation of the management system.

CASE STUDIES

FIRE

Steve is the training officer of a rural volunteer fire department. He must ensure all firefighters and officers have received the appropriate training to engage in departmental operations. Activities that members of the department must engage in range from managing the flow of traffic during an automobile accident to responsibilities that must be carried out by high-ranking officers.

SAFETY

Sue is faced with the daunting task of training a multi-generational workforce in all safety topics that pertain to their jobs. The plant contains production workers, a maintenance staff, and office workers who engage in administrative tasks, sales, and customer service. The facility is in the early stages of developing a safety program, and she is the first professional who has been hired to fill the role.

SECURITY

Sam is focusing on enhancing security training in the hospital at which he has been the Security Manager for ten years. He has held traditional classroom sessions for workers and management but believes other opportunities exist to enhance the level of awareness and importance of security in the hospital. His budget is somewhat limited, but he believes he can access resources throughout the hospital in human resources and information technology (IT) to develop a variety of innovative training formats.

- How can training be accomplished most effectively?
- What avenues of ongoing education can be used to support the communication of information throughout the year?

EXERCISES

For the following questions, identify a single facility environment in which you would like to situate your responses and answer each question accordingly. Answer each question in the context of a fire, safety, or security audit.

1. How can the learning styles of different generations impact the effectiveness of efforts in education and training?
2. What can be done to most effectively address the need of workers and managers to have control over their training?
3. What can be done to most effectively address the need of workers and managers to know what training content means to them in application to their work?
4. What problems exist when using PowerPoint presentations and videos as the sole methodology for delivering training?
5. What reasonable options are available to present options for training formats for initial training on necessary topics to support the implementation of the management system?
6. What training platforms are reasonable to use when developing ongoing customized training for workers and managers?
7. What regulations need to be explored to determine required training content, frequency, and trainer qualifications to address legal issues associated with the management system?
8. What ongoing education platforms are most useful for making workers and managers continually aware of management system information?

9 Evaluation and Continuous Improvement

The evaluation and continuous improvement components of a management system provide the opportunity to assess and review how well it is functioning. Quantitative and qualitative assessments are useful to collect data that can be analyzed to determine the effectiveness of the management system. The intent of this text is to explore this component of a management system in detail. This chapter will serve as a brief overview of information covered in more detail elsewhere in the text.

EVALUATION OPPORTUNITIES

Inspections and audits are the two primary tools used to assess the performance of a management system. Inspections tend to be designed to simply identify deficiencies. Items are listed on a form, and the inspector walks through a designated area to determine if each item complies or does not comply with a standard. Work orders can be generated by the inspector and submitted to the maintenance department to correct deficiencies. Examples of inspections include:

- Daily housekeeping inspections
- Pre-use forklift inspections
- Weekly fire inspections
- Monthly security inspection

Audits tend to result in quantitative measures that can provide an immediate perspective on how well a management system or a component of a management system is functioning. Audits can be designed to address a specific task or the measurement of the complete management system. Task-specific audits can be used to observe workers as their behavior is demonstrated to determine the degree to which the work is performed properly. Such audits can be voluntarily included as a component of the management system, or they might be required by a regulation. For example, an organization might voluntarily perform behavioral audits on workers to determine the degree to which work is performed safely, while OSHA's lockout/tagout standard requires periodic observations of workers performing lockout/tagout to ensure procedures are followed properly.

DELINEATION OF RESPONSIBILITIES

Responsibilities in inspection and audit processes should be delineated among management and workers. Responsibilities for management can include things such as the following:

DOI: 10.1201/9781003371465-11

- Provide training for inspectors and auditors to become competent in the use of the inspection or audit document
- Support the process by providing funding, correcting issues identified, and periodically communicating the measurements and benefits of the processes
- Conduct inspections and audits
- Provide time for workers to participate in inspection and audit processes

Responsibilities for workers can include things such as the following:

- Positively interact with inspectors or auditors
- Allow inspectors or auditors to observe work behavior
- Escort inspectors through the work area
- Receive training and become an inspector or auditor

ASSESSMENT

Inspections typically include simple responses to assess each item included on the inspection form. Assessment options include the following:

- "Yes" and "No"
- "Compliant" and "Non-Compliant"
- "Conformity" and "Non-Conformity"

The first response in each pair means the item is assessed as acceptable and the second response in each pair means the item is not acceptable. All items found to be not acceptable are then placed in a system for corrective action to address the issues associated with each finding.

Audits can include an assessment like that of inspections, where each item is simply identified as being acceptable or not acceptable. Behavioral audits are examples of where this methodology can be used to assess behavior and convert findings into a percentage score of proper performance. For example, one item on a worker's behavioral observation might address the need to wear safety glasses. Ten workers might be observed, with eight receiving a "yes" response due to wearing their safety glasses and two receiving a "no" response due to not wearing their safety glasses. The data can be converted to a score of the work being performed 80% safe due to eight of the ten workers wearing their safety glasses.

Management system audits can include a quantitative score to assess each question that is explored. For example, part of a safety management system audit might include assessing the confined space entry program. A question in the documentation review portion of the audit asks if entry permits are completed properly. Scoring options might include:

- 0: none of the permits were completed properly
- 2: some of the permits were completed properly
- 8: most of the permits were completed properly
- 10: all the permits were completed properly

This spectrum of scores provides the ability to reflect what is occurring compared to simple "yes" and "no" scoring options. For example, 20 permits might have been reviewed, and problems were found in three of them. A yes/no scoring paradigm would require a response of "no" due to existing issues, whereas a 0–2–8–10 scoring paradigm would result in an "8" being assessed due to most permits being completed properly. The 0–2–8–10 scoring paradigm provides the ability to assess scores that are more reflective of what is occurring in the workplace. The 0–2–8–10 paradigm also provides a metric that can be used to determine management system improvement or regression from year-to-year and can be used to compare multiple facilities within an organization. A management system with a yes/no paradigm will result only in a list of findings that must be addressed. The number of items on a list may or may not be reflective of improvement in the management system, making a yes/no paradigm less dependable in measuring management system performance over time.

ADDRESS FINDINGS

Findings from inspections and audits must be placed in a system for corrective action on deficiencies to occur. A basic system to prioritize findings to be addressed is:

- *High Priority*: fix immediately due to the high risk of injury or property damage
- *Moderate Priority*: fix soon due to moderate risk of injury or property damage
- *Low Priority*: fix once high and moderate risk findings have been addressed due to low risk of injury or property damage

Responsibility will need to be assigned to individuals to ensure each finding is corrected based on the timeframe established for each category. A follow-up process will need to be created to ensure each item is addressed.

REPORTING

The results of the inspection and audit process should be reported to key stakeholders in the organization. This will ensure awareness is maintained of the value of the processes and key findings of the processes. Agility can be used in considering organizational culture and norms to determine the format of the reports and the platform(s) of delivery. The audience for report distribution can be determined based on which members of management are primarily affected by an audit or inspection process. Certain lead workers or employee representatives should also be considered for report distribution.

UPPER MANAGEMENT REVIEW

Upper management should conduct an annual review of the management system. Subject matter experts, such as Safety Managers and Security Managers, can provide appropriate data for their review that has been derived from inspection and audit processes. Upper management can analyze the data and, through interaction with subject matter experts, determine what can be done to advance the performance of the management system most effectively.

CASE STUDIES

FIRE

Gene is the chief of her fire department and is concerned about the ability of the department to respond in the event of a hurricane due to the proximity of her city to the Gulf Coast. Though she has actively engaged in supporting the emergency response plans of facilities within her response area, she sees the need to turn the focus internally and begin assessing performance issues that might not have been previously considered. She believes the events of Hurricane Katrina sounded the alarm for such considerations due to the impact the hurricane had on the ability of public emergency services to respond.

SAFETY

Mary is the Regional Safety Manager for a manufacturing company that is building a new facility adjacent to the plant where her home office is located to expand operations. A plant Safety Manager has been hired to administer the safety management system at the new plant, and she is concerned the new hire is not working efficiently to implement existing safety programs in the plant opening process. The corporate office plans to conduct a management system audit at the new plant, but the intent is for it to be a "startup" audit that does not result in a score that is typically the result of management system audits throughout the organization.

SECURITY

Carl has tracked losses in his distribution center over the past year as a recently hired Security Manager. He has been able to use the data he has collected, but he believes he can elevate his efforts beyond simply analyzing incident reports and inventory reports. He would like to create a more robust system to get ahead of losses so that he can collect data that can be used to prevent incidents before they occur.

- How can inspections be used to improve the performance of the management system?
- How can audits be used to improve the performance of the management system?

EXERCISES

For the following questions, identify a single facility environment in which you would like to situate your responses and answer each question accordingly. Answer each question in the context of a fire, safety, or security audit.

1. What are the elements of the two primary evaluation tools?
2. What responsibilities should be delineated among workers and managers when using evaluation tools?
3. What assessment outcome options are available for inspections and audits? What limitations might exist in a yes/no paradigm?
4. What system should be put into place to ensure findings are properly prioritized?
5. What system of follow-up should be used to ensure all findings are brought to closure?
6. How should the results of inspections and audits be communicated throughout the organization? Who should receive reports?
7. Why is it important for upper management to conduct an annual review of the management system?

10 Host Employers, Contractors, and Staffing Agencies

Host employers, contractors, and temporary agencies are included within various sections of ANSI/ASSP Z10 and ISO 45001. OSHA created a unique area of focus on these relationships in its model. Making this a unique area of focus in a management system makes intuitive sense due to the significant level of risk that can exist when contractors and temporary workers are engaged in work at the site of a host employer. Contractors and temporary agencies might have a range of resources related to management system implementation, from no dedicated resources to subject matter experts in their employ. An analysis should be made by the host employer when engaging with contractors and temporary agencies to determine how to integrate them most effectively into the management system.

Four potential roles exist that should be considered. OSHA provides the following definitions to clarify these roles:

- *Host Employer*: An employer who has general supervisory authority over the worksite, including controlling the means and manner of work performed and having the power to correct safety and health hazards or require others to correct them.
- *Contractor*: An individual or firm that agrees to furnish materials or perform services at a specified price and controls the details of how the work will be performed and completed.
- *Staffing Agency*: A firm that provides temporary workers to host employers. A staffing agency hires its own employees and assigns them to support or supplement a client's workforce in situations involving employee absences, temporary skill shortages, seasonal workloads, and special projects.
- *Temporary Workers*: Workers hired and paid by a staffing agency and assigned to work for a host employer, whether or not the job is actually temporary.

The host employer, contractor, and staffing agency should communicate clearly from the start of the project and remain engaged throughout the coordination of the project. The host employer should assume a leadership role in directing this activity. OSHA states the following should be communicated prior to a contractor or staffing agency coming onsite:

- The types of hazards that may be present
- The procedures or measures they need to use to avoid or control their exposure to these hazards

DOI: 10.1201/9781003371465-12

- How to contact the host employer to report an injury, illness, or incident or if they have a safety concern

Similarly, OSHA states that individuals employed by the host employer should be aware of the following:

- The types of hazards that may arise from the work being done on site by workers employed by contractors or staffing agencies
- The procedures or measures needed to avoid or control exposure to these hazards
- How to contact the contract or staffing firm if they have a safety concern
- What to do in case of an emergency

Open and ongoing communication among all parties is needed for the management system to function at its full potential.

CONTRACTORS

Contractors may exist in three ways. First, contractors might be engaged to conduct limited work onsite, such as an electrician who has been contracted to repair a damaged electrical panel. Second, like the first category of contractors, some contractors might routinely visit the facility and provide periodic services. Third, contractors might be engaged in long-term activity onsite, such as a contract security company that manages access control to company grounds and buildings. A contractor management program should be established that directs how interaction with each of these types of contractors should occur. Components of such a program can include the following:

- *Prequalification*: Contractors can be selected based on reported safety performance in the contractor prequalification process. Requests for proposal (RFPs) can include a request for substantiated safety performance records, such as their insurance experience modification rate (EMR). An EMR assesses past workers' compensation history and estimates future potential performance. Proper insurance coverages should also be considered in the prequalification process, such as liability and workers' compensation.
- *Orientation*: An orientation process can be established to educate arriving contractors on the requirements with which they must comply in relation to the management system. Training is the responsibility of the contractor, and an orientation can be used by the host employer to translate how their previous training should be applied onsite.
- *Risk Assessment*: The host employer can make contractors aware of hazards they might encounter and how they are to be managed to avoid injury, fatality, and property damage. The host employer and contractors can partner in the assessment of emerging risks that might surface as a project unfolds.

- *Create Task Procedures*: Task procedures can be created and communicated to contractors. Procedures can include general operational, safety, security, and emergency response issues.
- *Observation and Follow-Up*: The host employer and contractor can partner to ensure all contract workers are following task procedures as designed. Periodic observations should occur to ensure task procedures are followed properly. A corrective action process should be in place to address nonconformance when performing tasks.
- *Ongoing Communication*: The host employer and contractor should engage in ongoing communications throughout the project to support the management system. For example, an avenue of open communication can be established that allows contractors to report hazards that are identified as the project unfolds. The contractor can also report safety performance issues, such as injuries and incidents that occur, as well as data from leading measures.

STAFFING AGENCIES

Staffing agencies may exist in two fashions in relation to oversite and management support. First, staffing agency management support might be limited to an offsite office in situations where few temporary workers are assigned to a facility. Second, staffing agency management support might exist onsite where a large volume of temporary workers is assigned to a facility. The ongoing presence of temporary workers who are directly engaged in production creates a unique dynamic in that they become an integral part of daily operations. Similar to contractors, a process can be implemented to safely integrate temporary workers into the workplace that includes the following:

- Orientation and training of temporary workers
- Risk assessment and communication of hazards to temporary workers
- Observation of temporary worker behavior and follow-up with the temporary worker and staffing agency personnel
- Ongoing communication with temporary workers and staffing agency personnel

The unique dynamic of the two-employer environment in which temporary workers engage has been identified as a concern and addressed by OSHA at https://www.osha.gov/temporaryworkers. OSHA provides detailed guidance on how to navigate such an environment to protect temporary workers. This guidance includes:

- Clearly define roles of each employer in the protection of the temporary worker.
- OSHA states, "…staffing agencies and host employers are *jointly responsible* for maintaining a safe work environment for temporary workers - including, for example, ensuring that OSHA's training, hazard communication, and recordkeeping requirements are fulfilled."

- "OSHA could hold both the host and temporary employers responsible for the violative condition(s) - and that can include lack of adequate training regarding workplace hazards. Temporary staffing agencies and host employers share control over the worker and are therefore jointly responsible for temporary workers' safety and health."
- *"Both* host employers and staffing agencies have roles in complying with workplace health and safety requirements and they *share* responsibility for ensuring worker safety and health."
- "A key concept is that each employer should consider the hazards it is in a *position* to *prevent and correct*, and in a position to *comply* with OSHA standards. For example: staffing agencies might provide general safety and health training, and host employers provide specific training tailored to the particular workplace equipment/hazards."
- "Staffing agencies have a duty to inquire into the conditions of their workers' assigned workplaces. They must ensure that they are sending workers to a safe workplace."
- "Ignorance of hazards is not an excuse."
- "Staffing agencies need not become experts on specific workplace hazards, but they should determine what conditions exist at their client (host) agencies, what hazards may be encountered, and how best to ensure protection for the temporary workers."
- "The staffing agency has the duty to inquire and *verify* that the host has fulfilled its responsibilities for a safe workplace."
- "Host employers *must treat temporary workers like any other workers* in terms of training and safety and health protections."

The host employer and staffing agency can partner to determine the risks associated with the work in which temporary workers will engage. They can then design a strategy to protect temporary workers most effectively against injury or fatality while performing work.

CASE STUDIES

FIRE

Jason works for the city and is responsible for managing preventive maintenance on all city-owned vehicles. His scope of responsibility includes fire apparatus at the three city fire stations. The city maintenance staff does not have the capacity to perform maintenance on all city-owned vehicles, so he decided to engage a contractor to perform maintenance on fire apparatus due to the unique nature of the equipment.

SAFETY

Phil is a new Safety Manager at a distribution center with a new safety management system. The facility routinely engages 40 temporary workers throughout the year. Temporary workers are assigned to the facility by the staffing agency; the workers arrive onsite, are greeted by their manager, and are placed on a production line. Phil is concerned about the safety risk to temporary workers and the lack of safety anywhere in the staffing agency relationship.

SECURITY

Clare is the Security Manager at a manufacturing plant. She worked with the existing contract security company to improve performance, with little response from the contractor. The contract period expires in 6 months, and she has decided to terminate the relationship and work to select a contractor that might serve as a more productive partner.

- What process should be used to select and integrate contractors in a manner that most effectively supports the management system?
- What ongoing avenues of communication should be established and carried out throughout the project?

EXERCISES

For the following questions, identify a single facility environment in which you would like to situate your responses and answer each question accordingly. Answer each question in the context of a fire, safety, or security audit.

1. In what ways might contractors interact with operations in the facility?
2. In what ways might temporary workers interact with operations in the facility?
3. What process should be implemented to analyze and communicate hazards to contractors?
4. What process should be implemented to prepare temporary workers for tasks in which they will engage?
5. What would be the characteristics of a successful relationship between a host employer and a staffing agency?

Part III

Audit Components

11 Documentation Review

Conducting the review of documentation should occur first among the three primary phases of conducting an audit, which are:

- Documentation Review
- Facility Inspection
- Employee Interviews

Conducting the documentation review first will allow the auditor to gain an understanding of what the facility communicates as being in place. The documentation review process will include the following:

- Determine the volume of records to review.
- Read written programs and policies.
- Evaluate supporting documentation that is used in the implementation of written programs and policies.
- Review training documentation.

VOLUME OF RECORDS

The documentation review portion of the audit will involve reviewing documents covering a broad scope of applications. The magnitude of these documents calls into question the volume of records that should be reviewed during an audit. For example, within the context of an annual safety audit, documents to review will include things such as:

- *Written Programs*: A facility could have 20 or more written programs when all compliance and non-compliance topics are taken into consideration.
- *Written Policies*: Written policies might be in place that act as general guidance with which the written programs are intended to comply.
- *Training Tests*: Tests may exist for each written program that serves as a tool to ensure employees understood the material that was covered in each training session.
- *Certificates of Completion*: A certificate of completion serves as an agreement between the trainer and employee that the list of information on the certificate was covered in the training session and the employee understood the material.
- *Training Logs*: A document for each training session might be used that lists the date of the training session, the name of the trainer, and the printed name and signature of each employee that was present.

DOI: 10.1201/9781003371465-14

- *Inspections*: Daily, weekly, monthly, quarterly, and annual inspection documents might be present that indicate periodic evaluations of things such as the sprinkler system and alarm system.
- *Permits*: Written permits might be utilized if welding has occurred in facility production areas or if entry has occurred into confined spaces that present hazards.
- *Injury Records*: A number of forms could be generated if the facility has experienced an injury, including an incident investigation form, a worker's compensation report of injury, medical records produced at the hospital where the injured employee was treated, an OSHA recordkeeping log, and an OSHA recordkeeping summary.

Each of these categories will result in a large volume of documents generated by the facility. The challenge is to determine the volume of these records the auditor should review. One option is to review all documents. This option is feasible at a facility that contains a small number of employees but may be a challenge at a facility that contains many employees. For example, a facility that employs 500 production workers would require the auditor to review many documents. This would require a great deal of time to be dedicated to reviewing training files alone. This issue causes a second option to be considered, which is to review a random sample of records that will provide an accurate representation of what is occurring at the facility. If random sampling is used, it is advisable for the auditor to select the percentage of files needed while onsite. If 500 employee files are present, a threshold might be to audit 10% of the files. Rather than ask facility personnel to have these records prepared in advance, the auditor can select the files onsite. This will avoid the risk of facility personnel selecting files that are known to be in complete compliance.

Where training records present a unique problem of sheer volume, it might be feasible to audit the remaining categories of records in their entirety. Each written program and policy will need to be reviewed for compliance. The volume of inspections should be in a quantity that will allow the auditor to review each one, with the exception of equipment pre-use inspections. All permits that have been issued should be able to be reviewed. Each injury file should be able to be reviewed.

This process will require a balance between the time it will take to review records and the risk involved in the documentation review process. The auditor must review a number of records that ensure an accurate assessment of what is occurring at the facility while performing the evaluation in an amount of time allotted for the audit.

WRITTEN PROGRAMS AND POLICIES

Written programs and policies serve as an indication of what is occurring at the facility. They document activity that a facility is engaged in regarding a given topic. Written policies are typically one or two pages in length and identify general issues that must be addressed in relation to a certain topic. Written programs are much more detailed and may range in length from a few pages to a larger program of more than 20 pages. Regarding a facility access control program, the following information might be found:

- Identification of the scope of activity
- Responsibilities of certain individuals
- Procedures used to enter and exit company property
- Procedures used to enter and exit buildings
- Procedures to enter and exit sensitive areas within the facility
- Procedures to manage contractors and visitors
- Equipment necessary to implement the program, such as keys, locks, and identification badges
- Employee training requirements
- Corrective action to be taken if individuals are found to be in non-compliance with the program.
- Revision history that chronicles changes made to the program
- Attachments of forms used to implement the program

The auditor will need to dedicate time to reading through each program, as these documents communicate compliance requirements and indicate what the facility is doing in relation to the topic covered by the program. This can be accomplished in two ways:

- *Prior to the Audit*: The auditor can request that the facility e-mail programs prior to the audit or access them through a shared drive or Internet platform. This will allow the auditor to conduct the program review prior to the audit. Reviewing programs in advance will maximize time onsite by dedicating all the auditor's efforts to those things that can only be accomplished at the facility. This will in turn save travel costs by minimizing the amount of time that is required to be onsite.
- *Onsite*: The program review can be conducted onsite. This will require a dedicated space where the auditor can work in relative quiet to read through the program material.

The documentation review portion of the audit may take as little as half a day to as much as a full day or more based on the scope of the programs being audited, the size of the facility, and the number of employees. Adequate time will need to be considered in the planning process to ensure sufficient time is given for this activity.

SUPPORTING DOCUMENTATION

Where written programs indicate what a facility communicates is being done, supporting documentation indicates how well the program is being implemented. This portion of the documentation review phase will typically be conducted onsite due to the handwritten nature of the documents. Though the forms will be well designed and printable, the individuals completing the forms will typically do so by handwriting on them through checking boxes or making notes. This will require the auditor to access hard copies of these completed documents while onsite. In some cases, documentation might be scanned and saved on an Internet platform that will allow the auditor to assess them prior to arriving onsite. Auditing these documents will require a few considerations:

- *Location*: Documents that support program implementation can be located in a number of places throughout the facility. Forklift pre-use inspections might be maintained in various departments where forklifts are used. Fire system inspections might be maintained in the maintenance department. Security system testing records might be maintained at the security base station or in the Security Manager's office. Due to the fragmented nature of where these records could be maintained, it will be necessary for the auditor to know where the records are in order to expedite the audit. Facility personnel could also gather these files in advance of the audit and have them ready in the auditor's workspace.
- *Content*: The auditor must be knowledgeable about the material being audited so that the forms can be properly interpreted. The auditor will be responsible for evaluating the supporting documentation to determine if compliance has been achieved and to provide feedback on how the forms are being completed or if other information on the forms would be beneficial.
- *Qualification*: Documentation used to support the implementation of a program must be completed by an individual who has been properly trained and is qualified to do so. For example, the person's name that is present on a quarterly fire system inspection form as being the inspector must be qualified to act in the capacity of an inspector. Similarly, a person who has completed a confined space entry permit must be qualified to effectively assess the level of safety prior to entry. The auditor will need to investigate the qualifications of these individuals to ensure they meet the requirements of being able to complete the forms that are on file at the facility.

The auditor will need to exercise attention to detail when reviewing supporting documentation for the implementation of programs. These forms are integral to the implementation process, and any errors that occur in their use must be identified and brought to the attention of facility personnel.

TRAINING DOCUMENTATION

Training documentation will represent evidence of communication of information, procedures, and personal responsibilities that correlate to material that is included in written programs. This can be present in the form of tests, certificates of completion, and training logs. The auditor will need to examine:

- *System*: The documentation system will need to be evaluated to determine if it is appropriate for the facility and legal requirements. The system will include the types of forms utilized to document training. A test may be utilized to ensure employees comprehend the information. Certificates of completion might be utilized to serve as verification that the employees agreed on the topics that were presented in the training session. A training log might be used as a record of who was present during a given training session. While some training records may be a matter of regulatory compliance, others may be a matter of compliance with company policy.

- *Content*: The training documents must contain sufficient content that accurately reflects the scope of material covered by the written program. Tests must cover the spectrum of information that is appropriate for the written program. Certificates of completion should accurately reflect the topics covered in the training session. Information on all training documents should be complete and legible.
- *Language*: Training documentation must take into consideration non-English-speaking employees that may work at the facility. If the auditor finds that there is a population of non-English-speaking employees at a facility, the auditor should find documentation that is in the employees' native language indicating that they understood the information presented in training sessions.
- *Trainer*: The trainer identified on the documentation must be qualified to conduct the training. The auditor will need to investigate the qualifications of the trainer to ensure that the individual is competent to conduct the training.

Documentation review can be the most tedious of the three phases of conducting an audit. It will require attention to detail and patience as each program and document is reviewed. The auditor will need to review the documents in their entirety, such as written programs, or may choose to review a sample, such as training files maintained by a large facility. A risk analysis will need to be performed to determine an acceptable number of documents to review in order to gain an understanding of what is occurring at the facility.

CASE STUDIES

FIRE

Adam is in the process of reviewing the training records as a component of his audit of Station 1. Everything appears to be in order, but then he notices the name of the trainer that is entered on the self-contained breathing apparatus (SCBA) training records. This training includes how to wear and maintain the SCBA as well as fit testing to ensure each firefighter can achieve a proper face-to-face piece seal. He is aware of the trainer but was not aware that he had gone through the training needed to conduct fit testing.

SAFETY

Amanda has asked the Maintenance Manager to provide her with all the confined space entry permits for entries that have occurred since the last audit. As she reviews each permit, she finds that they have all been completed properly. However, there appears to be a noticeable gap in the dates of permits. Permits seem to have been routinely issued throughout the year, except for two periods where there were periods of four weeks between permits being issued.

SECURITY

Don has reviewed both the corporate policy and the facility program on access control. The corporate policy mandates a Security Officer to be posted at the front gate that serves as the entrance to each facility. Rather than posting a Security Officer at the front gate, the facility utilizes electronic access that is monitored by a camera. An intercom is present in a visitor and contractor lane for them to contact the security office inside the facility to gain entrance. From the office, security personnel can view the individuals on the monitor and control the gate.

- What documentation review problems exist in these scenarios?
- How should the auditor proceed?

EXERCISES

For the following questions, identify a single facility environment in which you would like to situate your responses and answer each question accordingly. Answer each question in the context of a fire, safety, or security audit.

1. Should the review of documentation be the first phase of the audit process? Why or why not?
2. When can written programs be reviewed?
3. Is it acceptable to review only a sample of supporting documentation when conducting an audit? Why or why not?
4. How can the location of documents create a challenge during an audit?
5. What things should an auditor look for when auditing training documentation? What problems might be found?

12 Facility Inspection

The facility inspection can be conducted with knowledge of what was found during the documentation review. The facility inspection phase of auditing gives the auditor an opportunity to evaluate how well programs are being implemented in production areas. Whether it is a receiving dock or a fire ground, the auditor will be able to observe how well the implementation of information identified in the written programs is applied. Things to consider when conducting the facility inspection include:

- Inspect all areas
- Evaluate general conditions
- Observe employee behavior
- Inspect at various times

INSPECTION AREAS

The auditor will need to inspect each area of the facility. This will require a great deal of endurance due to the demands that will be placed on the auditor. The physical inspection may require walking for extended periods of time, climbing ladders, navigating confined areas of machinery, and being exposed to harsh climates.

An initial general walk-through of the facility will help to orient the auditor to the flow of production and the location of areas that will be of importance to the audit. Though someone will typically be available at the facility to guide the auditor on the inspection, it will be necessary for the auditor to ensure that every area within the facility is assessed. Areas that may be obvious will include:

- Production departments
- Office areas
- Break rooms
- Maintenance shop
- Exterior receiving and shipping docks
- Parking areas

The auditor will need to be attentive to possibilities for inspection that may not be readily presented. It may be clearly communicated to the person accompanying the auditor that all areas managed by the facility must be inspected. However, a tendency might be to only visit the primary production areas of the facility. This will require the auditor to be vigilant and ensure that all areas of the facility are visited during the inspection. In addition to commonly visited areas, the auditor will need to be aware of obscure and remote areas that must be inspected. This might include:

- Elevated areas that are obscured by machinery
- Basements

DOI: 10.1201/9781003371465-15

- Remote storage and electrical closets
- Fire pump rooms
- Outside buildings that are not in close proximity to the primary facility
- Operations that occur in buildings that are separate from the primary facility being audited.

The auditor will need to be alert to the areas that comprise the facility being audited. Questions may need to be asked of facility management to ensure that each area has been identified so that a comprehensive inspection can be performed.

GENERAL CONDITIONS

A primary goal of a facility inspection is to investigate the state of general conditions in the workplace. This refers to how the work environment appears at the time of the audit. Inspecting general conditions is a very simple process of walking through the facility and observing the conditions of the work environment in each area of the facility, but it requires the skill of critical observation. Auditors may not be subject matter experts on the operations conducted in the facility, but they should be very aware of the types of things that could cause an issue within the scope of the audit. For example, the auditor may not be immediately familiar with the production processes utilized to load and unload trucks, but the auditor can identify an electrical hazard that exists on an extendable conveyor belt that is used in the process.

The auditor will need to observe the various general conditions that exist throughout the facility. This might include things such as:

- Placement of fire extinguishers
- Storage of product
- Presence and location of eyewash stations and showers
- Presence and location of first aid kits
- Presence and location of automated external defibrillators (AED)
- Clearance of exit pathways
- Clearance of exit doors
- Clearance of sprinkler heads
- Clearance of space in front of all electrical panels
- Ergonomic condition of workspaces
- Functioning of emergency lighting
- Slip/trip/fall hazards
- Presence of hand rails on elevated work surfaces

Observing these and many other physical conditions in the workplace will help the auditor understand if what was stated in written programs is being implemented in the facility. By knowing what programs are in place and their content, the auditor can evaluate the degree to which issues in the written programs are being managed in the workplace.

EMPLOYEE BEHAVIOR

Though physical inspections are typically designed to evaluate general conditions, they should also be utilized as an opportunity to observe employee behavior while work is being performed. For example, if a security audit is being performed, the auditor can observe the behavior of Security Officers as employees, visitors, and contractors enter and exit the building. The written Access Control Program may indicate the Security Officers are to perform the following activities while at the building's primary entry point:

- Ask individuals to open all containers, such as backpacks, boxes, and brief-cases, for a visual inspection.
- Observe as employees scan their identification badge for entry.
- Ask visitors and contractors to wait in the lobby until their in-house point of contact can come to the front to greet them.
- Use the telephone, cell phone, or radio to immediately notify company personnel of the presence of a visitor or contractor.

Upon observing a Security Officer at the front entrance, the auditor may find that the personal bags of employees are not being checked. Backpacks are being taken out of the facility without being checked by the Security Officer. The auditor will interpret this as a finding in the audit due to the potential for employee theft.

Similarly, in a safety audit, the auditor may observe a work environment where lockouts are occurring due to the performance of maintenance on a piece of equipment. Having reviewed the written Lockout/Tagout program and being aware of legal requirements, the auditor is aware that the following steps must be taken:

- Production employees affected by the lockout must be notified that work will be occurring.
- The equipment will be turned off at a primary disconnect.
- Each maintenance employee working in the area will apply their lock to the disconnect to ensure that the equipment cannot be unintentionally started while work is being performed.
- The on/off switch will be tested to ensure that the power has been disconnected.
- Repairs will be made.
- Tools will be removed.
- Locks will be removed.
- The equipment will be energized to ensure it is operating properly.
- Production employees affected by the lockout will be notified that the equipment has been restored to operational capacity.

While conducting the facility inspection, the auditor observes the work being performed and takes advantage of the opportunity to determine if work is being performed in compliance with the written Lockout/Tagout Program and OSHA regulations. She notices that there are four maintenance employees working on the piece of equipment, but there are only three locks that have been applied to the point where energy is disconnected. This will be identified as a finding in the audit due to one employee not being properly protected by using a lock that is under the control of that individual. The auditor will also use proper channels to stop the work until the issue is resolved in this situation due to the immediate danger to which the maintenance employee is exposed.

Not all employee behavior can be observed during a physical inspection. In the examples previously mentioned, the auditor will need to be attentive to what is occurring in the work environment throughout the course of the physical inspection so that opportunities can be taken advantage of to determine if employee behavior is in compliance with established standards as indicated in written programs and mandated by federal, state, and local laws. If the work is not occurring, then the behavior cannot be observed. For example, the facility may have a confined space entry program. If a confined space entry does not happen to occur during the audit, then the auditor is unable to observe employee behavior. Staging an entry may not prove useful, as the auditor will not see an entry as it is being naturally performed. The lack of ability to observe an entry can be addressed by asking employees who are engaged in confined space entry questions during the employee interview phase of the audit.

INSPECTION TIMES

The physical inspection should occur at various times to ensure that a comprehensive perspective of what is occurring at the facility can be established. Most of a facility inspection can be performed by walking through the facility in one continuous session. The issue with performing that activity is that it provides the auditor with only one perspective on how the facility operates and the activities being performed during that shift. Though audit activity typically occurs during first shift hours, the facility inspection will need to be supplemented with a review of how things occur on other shifts.

In the event of a security audit, an inspection of the property perimeter may yield very positive results. It may be found that:

- Access control at the front gate is effective.
- Access control at the truck shipping and delivery gate is effective.
- Fencing is in good condition.
- Brush is sufficiently cleared away to allow a clear line of sight between the building and the fence.
- Cameras adequately cover sensitive areas.

The results of this portion of the physical inspection may indicate that the facility is performing at an exceptional level. However, a brief walkthrough of the area after sunset reveals that the lighting is insufficient. A number of the parking lights are out, and lighting is not provided in some areas covered by cameras, which provides those areas of camera coverage with no value after sunset.

Inspecting the facility during the second and third shifts may also reveal a difference in employee behavior. Where a lockout problem was identified in the previous example, it might be found that another shift is very diligent regarding the use of lockout procedures during maintenance activities. The finding then becomes an issue of how the facility needs to ensure a degree of consistency among all shifts when implementing programs, rather than it being communicated in the audit that there is a uniform problem with the implementation of lockout procedures.

CASE STUDIES

FIRE

Jason has completed the physical inspection of Station 3. He noticed at the end of the shift that the crew did not perform station cleaning as was designated in the housekeeping program. The audit would not conclude until the following day, so he decides to observe what type of housekeeping activity occurs during the next shift. Jason observes full compliance with the housekeeping program. The bay is swept, boxes of supplies have been received and put away, and all trash has been removed from the station.

SAFETY

Anne is running out of time to complete her audit. The documentation review portion of the audit took longer to complete than she planned. She wants to reserve enough time to complete the facility inspection. She only makes it through the primary production areas by the end of the day. She now only has until noon the following day to conduct employee interviews, prepare for and conduct the closing conference, and make it to the airport for her return flight home.

SECURITY

Terri is conducting the facility inspection when she observes a Security Officer reading a book while sitting at an assigned post. The post is a remote location of the facility that is used to monitor the transfer of high-value merchandise from its storage location into boxes to be shipped based on customer orders. This activity occurs only periodically throughout the day. Written post orders that Terri reviewed indicated that Security Officers are not permitted to engage in leisure activities while at their assigned post.

- How should the auditor manage what is presented in these scenarios?
- What principles of conducting an audit did the auditor violate or execute well?

EXERCISES

For the following questions, identify a single facility environment in which
you would like to situate your responses and answer each question accordingly.
Answer each question in the context of a fire, safety, or security audit.

1. What areas of a facility should be observed during a physical
   inspection?
2. What areas might present challenges when conducting a physical
   inspection?
3. What is meant by inspecting "general conditions"? What general
   condition issues might exist in an environment in which you would
   design and conduct an audit?
4. Is it important to observe employee behavior during a physical inspection? Why or why not?
5. At what time(s) would you conduct the physical inspection?

13 Employee Interviews

Conducting employee interviews is an opportunity to draw complete conclusions in relation to how the facility is performing in the areas that are within the scope of the audit. The documentation review phase gives the auditor an opportunity to evaluate written programs and supporting documentation. The facility inspection is an opportunity to observe how well programs are being implemented in the facility. Conducting interviews is the final component that allows the auditor the opportunity to determine how well program information has been communicated to employees and to what degree employees have internalized the information. Below are issues to consider when conducting employee interviews:

- Location where the interviews will be conducted
- Use of communication skills
- Types of questions to be asked
- Language barriers that may exist
- Confidentiality of those who are interviewed

LOCATION

Employee interviews will typically occur on the production floor or in a quiet area away from the noise and activity of work areas. Employee comfort is the primary factor in determining where interviews should be conducted. The auditor might be a person with whom employees are not familiar, thus making the interaction potentially uncomfortable or intimidating. Due to this potential dynamic, the employees should have input as to where the interviews are conducted.

Interviews conducted on the production floor will prove to be the most challenging for the auditor. First, the production floor might present a great deal of noise. This will cause a communication barrier between the auditor and the employee. Both may have to speak loudly and clearly to be understood. Though this barrier exists, a positive aspect of conducting interviews in such an area is that employees can immediately direct the auditor to a practical example of something that was contained in their response. For example, if an employee is asked a question regarding fire protection, the employee can demonstrate their knowledge of the topic by pointing out the location of fire extinguishers, sprinkler heads, alarm pull stations, and the path that they would take to exit the work area during an evacuation.

A second barrier that could exist when conducting interviews on the production floor is the potential for an employee to become distracted by work activity during the interview. Employees may feel more comfortable being interviewed in their work area, but it must be clearly understood by the Department Supervisor and the employees that the interview is to be conducted during dedicated time in which the employee will not be engaged in standard work activity. The employees

DOI: 10.1201/9781003371465-16

will need to remain focused on the questions being asked by the auditor without the distraction of attending to work duties. The auditor might need to address the temptation of employees to become visually distracted by work that is occurring in the department. This can be managed by the auditor positioning employees so that their line of sight is directed away from the activity that is occurring in the department.

Interviews conducted away from the work environment will be much easier to manage. The environment will be quiet, which will facilitate communication, and distractions in the work environment will be eliminated. Employees and the auditor will be able to solely focus on the questions being asked and subsequent interactions. Areas to consider for conducting interviews might include:

- *Lunch or Break Room*: A lunch or break room could provide a quiet and semi-private space to conduct employee interviews during periods when they are not in use. Attempting to conduct an interview in a lunchroom during a lunch break will create barriers similar to those that are present when conducting interviews on the production floor.
- *Private Office*: An office could be used temporarily to conduct the interviews. The furniture that is present in the room would need to be considered when deciding if this environment is to be used. A desk and visitor chair could create an adversarial environment where the auditor sits behind an expansive desk and the employee sits in the visitor chair on the opposite side of the desk from the auditor. A small task or conference table would be a better choice to use if present in the office. This would place the auditor and employee on an even level in that both are seated in similar chairs with no readily identified position of power at the table. It becomes an environment where a simple conversation can occur.
- *Auditor Work Area*: The area that has been temporarily assigned to the auditor as a work area might serve as a place where interviews can be conducted. The auditor will need to ensure that the area provides a degree of privacy and confidentiality where employees can feel free to answer all questions truthfully without fear of being heard by someone in an adjoining area.
- *Conference Room*: Conference rooms are an option for conducting employee interviews in that they are usually designed with privacy in mind. A door can be closed to prevent those in nearby work areas from hearing what is said inside. The conference table presents an opportunity for the auditor and employee to sit in positions with no readily identifiable position of power. The auditor will need to be sensitive as to where to sit in proximity to the employee. Sitting across the table from the employee might make the employee feel too separated from the auditor. The auditor will need to sit in a position that provides sufficient distance from employees so they feel comfortable, yet is close enough to create a sense of casual conversation.
- *Maintenance Shop*: Though the maintenance shop is considered a work area, it can be an area that is quiet and separated enough to facilitate

confidential employee interviews. This environment can serve to meet the need for maintenance employees to be interviewed in their work area, but it also serves as an effective environment for the auditor to conduct interviews.

Numerous options are available for conducting employee interviews. The auditor needs to select the environment that provides the greatest level of comfort for the employees. Though the area selected might be the production floor, which comes with unique challenges, the auditor can remain sensitive to the needs of the employee and overcome communication barriers to conduct an effective interview.

COMMUNICATION SKILLS

The auditor could be perceived by the employee as someone who is in a position of power, which could create an environment of inhibition. The auditor will need to exercise communication skills that place employees at ease. First, the auditor can speak "with" the employees rather than "to" them. This requires the auditor to create a relationship of equality with the employees. Though the auditor should be the subject matter expert, the knowledge and experiences held by employees can serve as a learning opportunity for the auditor. Both the auditor and employees have the opportunity to learn from the interaction.

Listening is a second skill that the auditor can utilize. Rather than dominating the interview, the auditor should simply state a question and then allow the employee to respond without direction from the auditor. Though it may be difficult to experience periods of silence, the auditor should allow time for employees to process the question and fully respond based on their knowledge and skills.

Third, the auditor should avoid being condescending. The auditor will be the subject matter expert, but the goal of the interview is to gain an objective understanding of what employees know and express their ability to exercise. The auditor should not critique employees' responses or make comments regarding the quality of training that the employee has received during the interview. The auditor should remain focused on speaking with the employee through a series of predesigned questions to gain an understanding of their knowledge and skills.

TYPES OF QUESTIONS

The auditor will need to enter the interview prepared with a series of scripted questions. These questions will be designed to address the scope of employee skills that are required among the programs that are addressed in the audit. These questions should be created as a standard component of the audit document to ensure consistency among audits. Audit questions must avoid a format that will result in a "yes" or "no" response. Examples of such questions are:

- Have you received training in Hazard Communication?
- Do you know where your assembly area is in case you need to evacuate during a fire?

- Do you complete a hot work permit prior to welding on the production floor?

Questions that result in a "yes" or "no" response will limit the auditor's ability to gain a deep understanding of what employees know and can exercise. Such questions will also lack the encouragement of employees to elaborate on the topic of the discussion. These questions can be better phrased as open-ended questions:

- What information was covered in your Hazard Communication training?
- Where is your assembly area in case you need to evacuate during a fire?
- What process occurs prior to welding on the production floor?

Open-ended questions provide employees with an opportunity to explain how an activity is performed rather than simply stating that it is or is not being done when a yes/no question is posed. Employees can more fully explain things, such as how work is conducted or the content of training they received.

If the response given by employees to the scripted question is lacking in information, the auditor can continue with probing questions. Probing questions are designed to explore the knowledge of an employee more fully by adding specific details that can be clarified. For example, an employee might be asked a question regarding the storage of hazardous chemicals in the facility. The employee responds with safe storage information but seems a little unclear about emergency response procedures should a spill occur. The auditor can then ask a probing question regarding access to safety data sheets and the information that is provided. This probing question allows the employee to communicate their knowledge more fully regarding the topic.

LANGUAGE BARRIERS

Language barriers in the workplace can exist in both rural and urban areas. Auditors need to be prepared for this potential and ensure that non-English-speaking employees are included in the interview process. Such employees must receive the same amount of information and training as English-speaking-employees, so it will be necessary to ensure this has been accomplished. An interpreter may be needed to facilitate the interview. Discretion will need to be exercised in choosing an interpreter. Though a bi-lingual co-worker who typically interprets for non-English speaking employees might be present, the audit program may dictate the use of a third party to ensure objectivity and confidentiality of the interview.

Hearing-impaired employees may also be present at the facility. This could require the use of sign language to effectively interview these employees. The word "sign language" is a little misleading because it is used in the singular tense when there are three versions of which the auditor must be aware:

- *Signing Exact English (SEE)* – SEE is a version of sign language that utilizes hand signs that exactly replicate all words and tenses of the English language. For example, in SEE, there are two signs that must be used when saying the word "running". One sign represents the word "run", and a second sign represents "ing", to indicate tense.

- *American Sign Language (ASL)* – ASL is recognized as an independent language. ASL is much more picture-oriented than SEE. Where SEE has a sign that replicates every word and tense that is spoken, ASL typically utilizes fewer signs to communicate a message than the words that are spoken. For example, the sentence, "I am going to the store." Would be stated in ASL with signs that represent only the words "I", "go", and "store".
- *Pigeon Signed English (PSE)* – Pigeon is a combination of SEE and ASL, where elements of each are represented. More signs may be used to represent spoken words in each sentence, but the signs representing tense are typically eliminated.

If there are hearing-impaired employees present in the workplace, the auditor will need to research the version of sign language that is spoken to select the appropriate interpreter for the interviews. The auditor will conduct the interview as with any other employee.

During the interview of a hearing-impaired or non-English-speaking employee, the auditor will make eye contact with the employee, ask the question, and maintain eye contact while the response is given. Even though the auditor does not know the language, he/she will maintain eye contact with the employee instead of the interpreter. It must be kept in mind that the interaction is between the auditor and the employee. The interpreter is merely there to facilitate communication.

Technology has greatly facilitated access to translation during interviews. Options include:

- Texting employees provides a platform for communication that allows the auditor and employee to type questions, responses, and follow-up interaction
- Platforms such as FaceTime and Zoom provide a video format to conduct interviews through an interpreter or translator

CONFIDENTIALITY

Employees must be assured that a degree of confidentiality will exist. Though their responses to questions will be used as a component of completing the audit document and will be shared in the audit report, the individuals who made certain comments will remain confidential. This level of confidentiality should help employees feel free to truthfully answer questions without fear of being identified.

Facility management must be informed that interviews are confidential and that employees must be free to share information as is appropriate when responding to questions. Though management may make assumptions as to which employees made various responses, as shown in the audit report, this cannot result in any type of retaliation against the employee. The integrity of the audit program is partially dependent on employees freely answering questions in a truthful manner. Fear of retribution may prevent authentic answers from being given.

CASE STUDIES

FIRE

Martha has planned with the Chief to conduct interviews in the break room. He has ensured Martha that only she and those being interviewed will be in the area, so no one will overhear the interview. Her first interview is with Sue, a second-year firefighter and the first female to be hired by the department. As Martha begins the interview, she explains that everything Sue says will appear on the final audit report, but no one will be aware of who made individual comments. Sue responds that she feels a little uncomfortable answering questions.

SAFETY

Elizabeth works in a department that is very loud due to the operation of machinery in the production process. Facility management has indicated that employees will feel more comfortable being interviewed on the production floor. When the auditor, Kim, arrives in the work area, she is aware of the challenge the noise will create. She is also confident that she can manage through it. She introduces herself to Elizabeth and begins the interview. After the second question, Kim notices that Elizabeth's attention is continually drawn to the operation of the machine to which she is assigned.

SECURITY

Jill has completed the documentation review and facility inspection and is now awaiting the arrival of her first employee to be interviewed. The individual is Sam, and his role is to monitor the images captured by the facility's surveillance cameras. Sam arrives on time and politely greets Jill. Jill starts the interview and soon realizes that whenever she speaks, Sam focuses intently on her lips. She realizes that Sam is depending on lip reading to communicate with her. Moments after she realizes this, she asks Sam if he is hearing impaired, and he responds that he is 90% deaf in both ears but can communicate well through lip-reading.

- Should the interview continue? Why or why not?
- What learning experiences can the auditor take from these experiences and apply to future audits?

EXERCISES

For the following questions, identify a single facility environment in which you would like to situate your responses and answer each question accordingly. Answer each question in the context of a fire, safety, or security audit.

1. Should the auditor be concerned about whether interviews are to be conducted on the production floor or in an isolated area? Why or why not?
2. What communication skills are most important when conducting an interview?
3. How should interview questions be phrased?
4. What should an auditor do if it is found that non-English-speaking employees work at the facility?
5. In what way is an interview considered confidential?

Part IV

Audit Phases

14 Pre-Audit

Conducting an audit involves three distinct phases of activity, each of which presents value to the process as a whole. The three phases of an audit program are:

- Pre-Audit
- Audit
- Post-Audit

The pre-audit phase is one of planning as well as execution. A great deal of planning is involved in that details must be properly planned and managed to design the audit in a way that ensures a successful event. Execution involves actually accomplishing certain tasks prior to conducting an audit. The pre-audit activity will involve the following:

- Scheduling the audit
- Maintaining ongoing communications with the facility
- Determine the number of auditors.
- Reviewing the application of regulations and company policies
- Arrange travel for the audit.

There are other things that must be done prior to an initial audit conducted within a newly developed audit program, such as auditor selection and training. Those issues are addressed in other chapters within this text. This chapter will focus on tactical pre-audit events that will occur routinely prior to each audit.

SCHEDULE THE AUDIT

Each facility within the organization will need to be evaluated to determine when audits should be conducted. An annual schedule can be generated to give facilities as much notice as possible to prepare for the audit. A number of issues will be involved in scheduling facility audits. These will include:

- *Production Schedules*: Each facility, whether it is a governmental agency, a manufacturing plant, or a non-profit organization, will have production concerns. The term "production" may vary among these organizations, so it will be incumbent upon the auditor to become aware of periods of high activity at a facility to be audited. These times should potentially be avoided due to the lack of focus that might be given to the audit while it is being conducted. For example, difficulties might immediately surface when conducting an audit at a school in the early fall. The school year is beginning, and all faculty and staff are focused on ensuring that operational systems

DOI: 10.1201/9781003371465-18

are in place to get the students acclimated and the academic year off to a good start. The auditor may find it difficult to gain the attention of faculty and staff at different points while conducting the audit. Similarly, an auditor might find it difficult to conduct an audit during harvest at a grain elevator when everyone at the targeted facility is working significant amounts of overtime. The opposite can also be true in that an audit might be conducted at a time when there is little activity occurring, giving the auditor limited ability to observe work processes at the facility. Production schedules should be evaluated to determine the most opportune time to conduct audits when activity within the facility is naturally occurring but does not place a burden on facility management.

- *Holidays and Vacations*: Holidays and vacations of key plant personnel should be taken into consideration when initially planning audits. Conducting an audit in close proximity to a holiday could create unnecessary challenges due to management and employees being distracted by upcoming personal or family commitments. It will also be important to ensure that key personnel are not away on vacation during the audit. This may be an issue only during the scheduling of initial audits, while it will be expected that future audits will be conducted at a similar time and vacations will be scheduled to accommodate the planning of the audit.
- *Facility Management Schedules*: Facility managers might have organizational commitments that will cause them to be away from the facility. This might include things such as a professional development conference or a quarterly business review meeting at the corporate office. The auditor will need to be aware of this potential so that scheduling conflicts can be avoided and all members of management who need to be present can attend the appropriate audit activities.
- *Time between Audits*: Where the previous issues take facility management into consideration, auditors will also need to pay close attention to their personal schedule to allow enough time between audits for the completion of administrative tasks. An interval between audits might be identified as two weeks. This will allow the auditor one day of travel to the facility, three days to conduct the audit, and one day of return travel in the first week. The second week can be dedicated to completing administrative tasks such as completing the audit report, distributing the report, responding to e-mails from the audit team supervisor, facility management, and others, and taking care of final details for the audit to be conducted the following week.

A schedule of audits can be developed that meets the needs of both the auditor and facility management. Considering production schedules, times when key personnel might be away from the facility, and appropriate intervals between audits will allow a schedule to be created that is acceptable to facility personnel and the auditor.

Rather than scheduling audits, some organizations prefer a format where audits are unannounced. This is a viable option, but it should be selected based on how well unannounced audits will be perceived among facility leaders. The rationale for an unannounced audit is that it will be conducted at a time when the facility is operating at a normal level. However, the audit will demand a great deal of facility resources, such as:

- Time of the facility manager to attend meetings
- Member of management to accompany the auditor on the facility inspection
- Time of the facility Safety Manager, Security Manager, or other responsible person to be always available
- Time for employees to engage in interviews conducted by the auditor

The decision to conduct unannounced audits should be made with these resource requirements in mind. Production demands and other variables may make it difficult for facility management to comply with all resource requirements during an unannounced audit. Unannounced audits can be more effectively achieved if the requirements of the audit program are communicated with sufficient advance notice that will allow all facilities to prepare and have a contingency plan in place to supply the needs of the auditor.

COMMUNICATIONS

The auditor will need to engage in ongoing communications with facility management from the time the audit is scheduled until the day the audit is conducted. The initial communication will be to inform the facility of the date the audit will be conducted. The auditor should request confirmation that this date is agreeable.

Periodic communication should then occur to remain in contact with facility management. This contact will ensure that communications regarding the audit are consistently understood. It will also provide location management with the opportunity to ask the auditor questions regarding compliance issues or the audit process.

The auditor should communicate what facility management should expect as they experience the audit. They should be made aware of what resources will be needed during the audit, including physical and personnel resources. Physical resources will include things like a place for the auditor to work. Personnel resources will include things such as arranging for someone to accompany the auditor during the facility inspection and the need to spend time interviewing employees.

The auditor should also communicate what the facility can do to most effectively prepare for the audit. One tool that can be used is to develop a checklist that can be e-mailed to facility management and that includes all items that the facility can use to prepare for the audit. This document might include sections for:

- *Logistical Considerations*: The document can itemize logistical needs for the audit. This will include a place for the auditor to work, a place to interview employees, and a location to conduct opening and closing conferences. The auditor can specify preferences for each logistical item, such as the auditor's workspace being in a quiet area that provides a degree of privacy.
- *Personnel Needs*: Individuals that will be needed at various points during the audit can be delineated on the document. This might include the Facility Manager and others attending the opening and closing conferences,

the number of employees to be interviewed, and the facility point of contact accompanying the auditor during the majority of the audit.

- *Documentation to be Reviewed*: The auditor can delineate the documentation that will be reviewed on site during the audit. This can include things such as written programs, inspection forms, incident investigations, and fire system inspection records. Facility management can then collect these items prior to the audit, which will facilitate the process by having everything on hand when the auditor arrives.

The audit can become an efficient use of everyone's time if expectations are clearly understood by all who are involved. Facility personnel can utilize the information gained from communication to effectively prepare for the audit. This will allow the time of those at the facility and that of the auditor to be utilized efficiently. Rather than attempting to schedule employee interviews and collect documentation while the auditor is on site, prior communication can be utilized to expedite the process by being thoroughly prepared.

NUMBER OF AUDITORS

Each audit will need to be evaluated to determine the number of auditors that will be required. There are a number of variables to consider in order to make this determination. Influencing factors can include:

- *Scope of the Audit*: An audit that is designed to cover multiple disciplines may require the presence of independent auditors who have unique expertise in each area. For example, if the annual audit includes both safety and security within its scope, then a safety professional and a security professional may be needed to carry out the audit. In some cases, a single auditor with expertise in both disciplines may be appropriate.
- *Size of the Facility*: The size of the facility may dictate the number of auditors needed to accomplish an audit. A single building of 500,000 square feet may require only one auditor, while a campus containing ten separate buildings and multiple production processes may require two or more auditors.
- *Complexity of the Operation*: Though a facility may be relatively small, operations within the facility may be complex. This could require an excessive amount of time to audit each area, or it may require auditors with special expertise who can properly assess the operations.
- *Time Allotted for the Audit*: The amount of time that has been scheduled to conduct the audit can affect the number of auditors selected. A large facility with numerous production processes may only be allotted two days in which to complete the audit. This could require the use of multiple auditors to assume responsibility for various areas of the facility.

Each of these variables can greatly impact the number of auditors that will be needed to accomplish a thorough audit. In many cases, these variables will need to be assessed as a group to ensure that each aspect of the audit is addressed so that the appropriate number of auditors can be assigned to the task.

REGULATION AND COMPANY POLICY REVIEW

Auditors will need to ensure they are knowledgeable of all regulations and company policies that apply within the scope of the audit. Auditors will be considered subject matter experts on these issues and their application to the audit process. Auditors will need to be familiar with the application of regulations and company policies in order to properly assess each audit question. The auditors will need to determine if the condition of the facility:

- Meets the requirements for a given question
- Does not meet the requirements for a given question
- Does not apply to the given question

If the response to the question is that the facility does not meet the requirements or that the question is not applicable, the auditor must be able to articulate why either of these responses was given. Facility management will need to understand that effort does not need to be placed in that area if it is not applicable to them or what needs to be done to correct the situation if the auditor believes that the facility does not meet the requirements.

A unique situation may arise that renders an auditor unable to make a decision regarding a specific audit question. In such a situation, the auditor can inform facility management that the issue will be researched and a final decision will be communicated prior to the audit report being generated. It is unacceptable for the auditor to place the burden on facility personnel to conduct this research while indicating a negative response to the question in the audit based only on the auditor's belief that a problem exists. The auditor holds the responsibility of basing all audit responses on substantiated fact.

TRAVEL

Travel will need to be scheduled to accommodate the scheduled audit. Auditors will need to plan their travel so that they arrive in sufficient time to begin the audit. If an audit is scheduled to begin at 8:00 a.m. on Tuesday, the auditor should arrive on Monday evening in the event that a flight connection is not made, or mechanical problems occur with a rental car.

Auditors will need to comply with company travel policies. These policies may dictate issues related to each aspect of the auditor's travel, including:

- Plane ticket cost thresholds
- Lodging cost thresholds
- Preferred car rental companies
- Per diem amounts available for food and beverages while traveling
- Other reimbursable expenses
- Expense reimbursement procedures

Auditors will need to be aware of these policy requirements and ensure compliance. They will also need to ensure ethical conduct in making decisions and submitting expenses based on policy compliance and not personal gain.

CASE STUDIES

FIRE

An audit team has worked for a number of months to create the city fire department audit program and audit document. They are now ready to begin scheduling the audit for each of the city's five stations. Upon contacting Station 3, they speak to Chief Dawson, who is very excited about the audit process. She wants to make improvements in her station and hopes that the audit will help give her direction. They also find that she is scheduled to attend a national fire conference on two of the three days for which the audit is scheduled. She will be there on the first day of the audit but will be out of town the last two days of the audit. She would like to maintain the current schedule for the audit because rescheduling it would mean pushing it back two months.

SAFETY

Acme Manufacturing has had Plant 5 scheduled for its annual safety audit for six months. The auditor has remained in contact throughout the period leading up to the audit to help make sure the plant is prepared. Two weeks before the audit, the Plant Manager contacts the auditor to inform her that the schedule for the period of the audit has just become busier than originally planned. A new order of parts has come in, and the facility will be working overtime for the entire week. The Plant Manager believes the audit can still be conducted.

SECURITY

Acme Distribution will have its annual security audit performed in three weeks. The warehouse is protected by posts that are controlled by proprietary security officers. The facility Security Manager contacts the auditor to let him know that the Corporate Security Director just called to say that he will be in town and would like to conduct training with all Security Officers on a new program that will require three days. The three days are the same three days scheduled for the annual security audit. The corporate Security Director has made arrangements with a contract security company to provide personnel to cover each post during the period that the training will be taking place.

- Should the audit be done according to schedule?
- What options are available to the auditor?

EXERCISES

For the following questions, identify a single facility environment in which you would like to situate your responses and answer each question accordingly. Answer each question in the context of a fire, safety, or security audit.

1. What variables should you consider when scheduling audits?
2. What system of communications will you implement to help ensure the facility is prepared for the audit?
3. What variables should you consider when determining how many auditors will be needed for a given audit?
4. What degree of responsibility does the auditor hold in knowing the regulations and company policies that apply to the facility being audited?
5. What aspects of an organization's travel policy should be understood when engaging in audits?

15 Audit

The planning and preparation that were accomplished in the pre-audit phase will result in the auditor arriving at the facility on the appointed day to conduct the audit. Prior communication and audit scheduling will have set the stage for the arrival of the auditor. A great deal of work will have gone into the pre-audit phase that occurred over a period of months. The audit phase will be very active from a different perspective. Where the pre-audit phase involved smaller sets of activity covering a number of months, the audit will be a very concentrated period of activity that will occur over a few days. The auditor's full attention will be on the facility that is being visited. Activities that the auditor will accomplish during the audit will include managing the following:

- Opening conference
- Facility tour
- Documentation review
- Facility inspection
- Employee interviews
- Facility management updates
- Data entry into the audit document
- Training of location personnel
- Closing conference

OPENING CONFERENCE

An opening conference will be the first activity of the audit. This conference is an opportunity for the auditor to establish open communication with key stakeholders in the audit. These individuals might include:

- Facility Manager
- Shift Managers
- Safety Manager
- Security Manager
- Maintenance Manager
- Department Supervisors
- Employee Committee Officers
- Chief
- Training Officer

Those present in the opening conference should be all members of facility management who have responsibility for items that are within the scope of the audit. Though this list could become extensive when considering additional personnel at

DOI: 10.1201/9781003371465-19

the facility, the judgment of the auditor and communication between the auditor and the Facility Manager should be used to determine who should be present. Some opening conferences might involve only an executive leadership team of a few managers, while in other cases, they might involve all members of facility management. An opening conference with an executive leadership team will allow the auditor to address important issues on a more direct level with those who are primarily charged with making decisions that affect the facility. A large opening conference with all members of management present will allow the auditor access to individuals in all positions of leadership to communicate the importance of the audit and the value that it can have on daily production activity at the facility.

The decision of which format to use will depend on the organizational culture and the perspective of the Facility Manager. Organizational culture will influence who attends various meetings based on the degree of engagement and transparency of facility operations. Organizations that elicit the support and engagement of everyone might be more likely to have all members of management attend an opening conference so they can become directly aware of the audit and what will occur over the course of the days to follow. Organizations that operate in a hierarchical fashion may limit the opening conference to an executive leadership team. In other cases, it might be the simple preference of the Facility Manager or the auditor as to who attends. Fundamentally, members of management with responsibility for correcting audit deficiencies and managing the processes being audited should be present.

Employee committee officers could also be invited to attend the opening conference. For example, the chair of the Safety Committee could be invited to attend the opening conference of an annual safety audit. This will demonstrate the support for employee engagement.

The opening conference is an opportunity to set a positive tone for the audit. The auditor will need to be aware of the value of making a good first impression on those who are present. The opening conference might be the first time that the auditor has met certain members of management. This will be an opportunity to demonstrate competence and professional expertise in the area of the audit, which will in turn establish a degree of credibility among those who are present. Perceptions of the auditor can influence perceptions of the audit process. If the auditor makes a bad first impression and is not perceived as credible, personnel in the opening conference may begin to perceive a lack of value in the audit.

The auditor will need to address the purpose, flow, and resolution of the audit process so that everyone understands what to expect. This will include a discussion of:

- *Operational Hours of the Facility*: The auditor will need to confirm the operational hours of the facility and the personal work schedules of those involved with the audit. The auditor will need to accommodate facility and personal schedules when conducting an audit. For example, if a facility utilizes a predominantly first-shift operation that concludes at 5:00 pm, the auditor should plan to conclude work for the day at 4:30 pm and use the last 30 minutes of the day to bring that day's work to a close. Though the audit is a significant event, the auditor will need to work within the typical work schedule as much as is feasible to demonstrate respect for facility

personnel. Demanding that facility management stay late to accommodate the auditor could generate hostility toward the auditor and the audit
process.

- *Phases of the Audit*: Those who are present for the opening conference will
  need to be aware of the phases of the audit, which include documentation
  review, facility inspection, and employee interviews. This will help them
  be aware of approximately when they might need to engage in the audit,
  such as by providing employees to be interviewed. The audit will be an
  invasive process that will affect the ability of management to carry out
  its work. Employee interviews will be conducted that will take employees away from their work. A facility inspection will be done, which may
  require supervisors to accompany the auditor so that any questions can be
  promptly answered. The opening conference is an opportunity to address
  these events so they can be planned for by facility management.
- *Meeting Arrangements*: The auditor will need to provide periodic updates
  as the audit progresses. The way these updates will be provided will need
  to be determined at the opening conference. These updates may range from
  an informal meeting at the end of each day between the auditor and the
  Facility Manager to a formal meeting with key personnel.
- *Scoring of the Audit*: Facility personnel will be interested in how well they
  performed on the audit. The scoring methodology should be reviewed in
  the opening conference. The auditor should also communicate when the
  final score will be communicated. The auditor can provide a rough estimate in the closing conference with a final score to be communicated when
  the audit report is finalized and distributed.

The auditor should be well prepared when conducting the opening conference. It
is an opportunity to make a positive impression on those present and to review information that will ensure a successful process as the audit is conducted.

FACILITY TOUR

A facility tour will help to orient the auditor to the operations and unique activities
that will be occurring in the facility throughout the audit. A member of facility management will accompany the auditor to ensure safety. The member of management
will be aware of risks inherent in certain areas of the facility and can communicate
this information to the auditor.

This tour will help the auditor understand the flow of production in the facility
and the work being performed. This information will help to inform the understanding of the material that will be explored when performing the documentation review.
The tour will also alert the auditor to opportunities to perform workplace observations during the facility inspection.

The facility tour should not be confused with the facility inspection. The facility
tour is designed to be a concise walk-through and orientation for the auditor to the
facility. The facility inspection is designed to be a thorough analysis of each process
and every area of the facility in relation to established standards.

DOCUMENTATION REVIEW

The documentation review will be the first of the three core phases of an audit. Conducting the documentation review first among the three phases will help the auditor to create an understanding of what programs and procedures the facility has in place, which can be used to direct activity within the scope of the audit. For example, a fire station's Personal Protective Equipment Program will identify what equipment is used, where it is stored, and how firefighters are trained to use it. This foundational information will be necessary when conducting the facility inspection and employee interviews to determine if what is written in the documents is actually being carried out in practice.

The auditor will need to review a broad spectrum of documents to properly assess the level of compliance that a facility has achieved. The two types of documents to be reviewed are written programs and supporting documentation. Written programs articulate what the facility is doing to manage a specific issue. A facility's Confined Space Program will delineate things such as the following:

- Purpose of the program
- Scope of the program
- Delineation of responsibilities among management and employees
- Equipment needed to implement the program
- Procedures used to carry out the program in the workplace
- Employee training content and requirements
- Corrective action that is provided when individuals are found to be non-compliant with the program
- A revision history of changes made to the program

Supporting documentation includes all forms that are utilized to implement written programs. For example, a Forklift Program will require the use of a pre-use safety checklist that operators must complete. Supporting documentation will include things such as the following:

- Training Tests
- Certificates of Completion
- Training Logs
- Inspection forms
- Permits
- Injury Records

Documents such as these may be found at various places throughout the facility, including the maintenance shop, departments, and manager offices. Planning will need to occur to expedite the collection and review of these documents.

Some parts of the documentation review process might be achieved in the pre-audit part of the audit. Written material can be e-mail to the auditor or accessed through an online platform, such as Drop Box or SharePoint. Conducting as much of the documentation review in advance will assist in maximizing the use of on-site time in conducting the physical inspection and engaging with management and workers.

FACILITY INSPECTION

Where the facility tour serves as a brief orientation to the facility and its operations, the facility inspection is a thorough review of everything that is occurring in the facility. The auditor will inspect every area of the facility to determine the degree to which the material in the written programs is being implemented. To accomplish this, the auditor will explore every part of the facility.

The facility inspection will require a degree of physical endurance. The auditor can be exposed to things such as inclement conditions, climbing ladders, and extended periods of time spent walking. The auditor will need to take into consideration the physical stress that will be required. The stress of an audit may affect auditor assignment due to personal physical capabilities and the pace at which the inspection is conducted. Breaks can be utilized to provide an opportunity for rest and refreshment.

EMPLOYEE INTERVIEWS

The final phase of an audit will be to conduct employee interviews. Here, the auditor will take into consideration what was found during the documentation review and facility inspection to explore the knowledge of employees on various issues. Open-ended questions will be used to elicit substantive responses from employees. Such questions allow employees to explain information in depth. Follow-up questions can be used by the auditor to explore issues and clarify an employee's understanding of a given topic more deeply.

Employee interviews should be conducted at a location that is comfortable for them. Some employees may feel more comfortable responding to questions on the production floor in their natural work environment, whereas other employees may feel more comfortable in a quiet area that provides a greater degree of confidentiality. The auditor will need to be sensitive to where the employees would prefer to conduct the interviews and accommodate their needs.

MANAGEMENT UPDATES

An audit may take place over a period of days. Three workdays may elapse between the time of the opening conference to initiate the audit and the closing conference to conclude all audit activity. Periodic updates should be scheduled throughout this time to maintain management focus on the audit throughout the event. Management updates can be as informal as providing a five-minute briefing to the Facility Manager at the end of the day that highlights key findings and best practices. The updates can be as formal as a brief meeting with selected members of management at designated times. The format, attendees, and location of these meetings should be determined in the opening conference and managed by the auditor as the audit progresses.

DATA ENTRY

The audit will generate a large amount of information, to include:

- Simple responses to the audit question
- Notes the auditor will generate to clarify simple audit responses
- Notes the auditor will generate in relation to employee interview responses
- Digital photos to document a deficiency or best practice

Though time will be dedicated in the post-audit phase to complete the audit report, the auditor should input as much information as possible while the audit is occurring. Allowing a significant amount of time to pass between collecting the information during the audit and entering it into the document could increase the likelihood of an error occurring or forgetting what was meant by the notes that were taken. The auditor should schedule time each day to update information in the audit document while events and written notes are readily remembered.

TRAINING

Though the audit is primarily designed to identify opportunities for improvement and best practices, it can also be used as an opportunity to train individuals as the audit progresses. For example, a safety audit might be underway, and the facility Safety Manager is new to the profession. She has a degree in safety management but has little experience. The auditor can use the audit as an opportunity to help train the new Safety Manager on issues related to managing things that are within the scope of the audit.

CLOSING CONFERENCE

The closing conference will occur once the auditor has completed the documentation review, employee interviews, and facility inspection and has updated the audit document with as much data as possible. Those present at the closing conference should be the same individuals as were present at the opening conference. The auditor will review the following information:

- *Primary Findings*: The auditor will have found deficiencies that pose a low, moderate, or severe risk to the people and the facility. The auditor should highlight the findings that represent the greatest risk. Specific examples should be used to help facility management understand what has been identified. Though photos will be included in the final report, the auditor can explain exactly where certain issues existed, what the deficiency was, and what the facility can do to correct the problem.
- *Best Practices*: Conducting an audit is an opportunity to identify where things are going well in addition to finding opportunities for improvement. Best practices can be identified during the audit and highlighted in the closing conference to encourage facility management in areas where they are performing well.

- *Audit Score*: If the audit is scored, facility management will be very interested to learn how they performed. Though the audit report will take a number of days to generate, proofread, and finalize, a general estimate can be communicated. Facility management will need to realize that the score communicated in the closing conference is an estimate based on information that has been entered into the document at that point but may change slightly as the audit report is finalized.

The auditor should express appreciation for the time and effort that facility personnel put into preparing for and engaging in the audit. Facility management should also be informed as to what to expect in the near future, including when the audit report will be completed and how follow-up will occur in addressing the correction of audit deficiencies.

CASE STUDIES

FIRE

Jean is preparing to go into the closing conference with the chief and station officers after conducting the station's annual audit. In reviewing her information, she notices that she checked "No" in response to a Personal Protective Equipment Program problem that was related to bunker gear. However, after reading through all her notes and reviewing all of the digital photos, she cannot remember the nature of the problem. The closing conference is in 15 minutes.

SAFETY

Luke arrived at the facility to conduct the annual safety audit that had been scheduled for six months. The opening conference is scheduled for 8:00 am, and Luke is expecting six members of management and the chair of the Safety Committee. At 8:15 am, only two members of management and the Safety Committee Chair are present. Luke has been notified that the remainder of the managers, including the Facility Manager, will be in another meeting until 11:00 am.

SECURITY

Donna is conducting the opening conference for the facility's annual security audit. She begins to address the need to provide periodic updates to management as the audit progresses. The Facility Manager responds by saying that the updates are unnecessary. He continues by saying that the facility has just received a substantial product order, and everyone will be focused on getting the order fulfilled over the course of the next three days. He feels that the closing conference will be sufficient.

- What should the auditors do?
- What can be done to correct these problems in future audits?

EXERCISES

For the following questions, identify a single facility environment in which you would like to situate your responses and answer each question accordingly. Answer each question in the context of a fire, safety, or security audit.

1. What information should be reviewed in the opening conference? Who should be present?
2. How does a facility tour differ from a facility inspection?
3. What items are examined during the documentation review phase of the audit?
4. What things should be taken into consideration when scheduling employee interviews?
5. Why are periodic updates of progress through the audit important?
6. When should data be entered into the audit document? Why?
7. How can training occur during an audit?
8. What information should be reviewed in the closing conference? Who should be present?

16 Post-Audit

A great deal of energy will have been expended in the pre-audit and audit phases. Though there is a sense of accomplishment in completing an audit at a facility, work remains to be done in the post-audit phase to bring the process to closure. Activities that the auditor will need to engage in during the post-audit phase include:

- Complete the population of the audit document with comments and photos
- Proofread the audit document
- Distribute the audit report
- Follow up with the location on correcting deficiencies

COMPLETE THE AUDIT DOCUMENT

The audit document should have a great deal of information input into it by the time the onsite audit activity is complete. However, there will typically be additional information that needs to be added once the auditor returns to the office. This might include:

- *Final Assessments*: Questions may have arisen during the audit regarding the application of regulatory or company policy issues that the auditor could not readily assess at the time of the audit. This may require the auditor to perform additional research upon returning to the office. Once the information is identified and a decision is made, the auditor can update the audit document with a final assessment of the audit question. Though this information will appear on the final report, the auditor should immediately notify facility management of the decision so that final conversations can be had prior to the audit report being finalized.
- *Comments*: The auditor will take a lot of notes throughout the site visit. During the visit, the auditor may only have time to enter the fundamental response to each question, such as "yes," "no," or "N/A," while entering basic comments into the audit document. The auditor will have time following the audit to read through the notes that were taken and enter more robust information that provides detail regarding findings and best practices. The auditor should strive to conduct as much of this activity during the site visit while ideas are readily remembered. There will typically be limited time to do so, causing some of this work to be performed upon returning to the office at the close of the audit.
- *Photos*: Digital photos are a tool that can be used to specifically direct facility management to findings during the audit. They can also be used to record evidence of best practices to be communicated to other facilities. The photos taken during the audit will need to be downloaded and entered into the audit document in the exact area to which the photo applies.

DOI: 10.1201/9781003371465-20

Completing the audit should be accomplished as soon as possible upon returning to the office. Memory will begin to fade as each day passes, so it is important to make the completion of the report a priority. The form should be completed in no more than one week upon returning to the office. The more time that passes brings new disruptions and activities that can have a negative impact on the quality of the final information that is recorded in the audit document. Completing the form as quickly as possible will help to ensure that each observation, note made, and photo taken by the auditor will be accurately reflected and explained in the audit document.

PROOFREAD THE AUDIT DOCUMENT

The audit document will need to be proofread to ensure that a professional report will be provided to each person who will receive the audit report. Information will be entered into the audit document over a period of weeks and under a range of circumstances, from the quiet of an auditor's personal office to the noise and activity of the facility being audited. Errors can easily be made when entering information. Errors that can occur include:

- *Dual Entry of Question Responses*: Each question in the audit can only have one response. It will either be "yes," "no," or "not applicable." The document will need to be reviewed to ensure that no dual entries were made for questions throughout the audit. The entry of dual responses will cause the scoring calculation to present inaccurate data.
- *Manually Entering Data*: The audit document should be created in a way that utilizes formulas to calculate scores for sections and the audit as a whole. The auditor will need to be aware of these cells and not attempt to manually enter data in them. Manually entering data will erase the formula. If data is manually entered, erasing the formula, changes in question responses that might be made later will not automatically be reflected due to the formula being replaced by a manually entered number. These cells can be protected, so the auditor is unable to enter data into these areas of the audit document.
- *Alignment of Photos and Findings*: A large number of photos might be taken during the course of an audit. A method will need to be established to track each photo taken to ensure it is properly aligned with an audit question and comments in the final audit report.
- *Cover Page Information*: The cover page of the audit document will contain information that might be similar among audits. This might cause the auditor to use the "save as" or "copy and paste" functions in the software to avoid retyping the information. If this is done, the auditor will need to proofread the material to ensure it has been customized to apply to the facility for which the report is being made.
- *Grammar and Spelling*: The audit document and report should be presented using technical writing skills. Sentences should be well structured and clearly articulate what the auditor is attempting to communicate. All words within the document should be spelled and utilized correctly.

Spelling can be addressed through the software's "spell check" function. The auditor will need to proofread the document to ensure proper word choice, such as in the use of "to," "too," and "two." Spell check will accept each of these three words, so the auditor will need to ensure the proper word was used.

Proofreading the audit document can be a mundane process, but it is critical to delivering a quality product to upper management and facility management. The credibility of the audit program can become suspect when numerous errors are found in an audit report. Proofreading and making appropriate corrections will result in a quality document that provides information on which those reading it can depend.

DISTRIBUTE THE AUDIT REPORT

The audit report can be distributed once all the information has been entered and it has been proofread. The audit program may dictate the distribution list for audit reports. One concern that must be taken into consideration in distributing the audit report is the legal protection of the audits. Court proceedings in the event of a lawsuit can cause various company documents to be discoverable evidence. The organization will need to evaluate if audit documents are to be protected or are to be open to discovery in the event of a lawsuit. Fundamentally, there are two decisions that can be made:

- *Unprotected*: The organization believes that broad internal knowledge of the audit report outweighs the limitations that will be imposed if the report is to be protected. Information is openly shared so that the organization can learn from the audit performed at a given facility.
- *Protected*: The organization believes the audit report is sensitive and should be protected from discovery. For example, all audit reports might be directed through the office of in-house legal counsel, who then manages the communication of information from the audit as attorney-client privilege. This process will typically result in fewer people within the organization receiving the audit report.

The degree of protection afforded by an audit report can have an impact on a distribution list. If the audit report is considered unprotected, it can simply be attached to an e-mail and distributed to a wide range of managers within the organization. If the audit report is protected, strict guidelines will be established for the distribution of the report through the organization's legal counsel.

The distribution list for an audit report can be created once it has been determined if the audit will be protected or unprotected. Individuals to consider including on the distribution list are the following:

- *Vice President of Operations*: The person responsible for the operations of a division within an organization will be able to read the report and hold those within the division accountable for performance on the audit.

- *Regional Manager*: The person responsible for a number of facilities will be able to hold the location accountable for audit performance as well as work with location management to provide the resources to address findings in the audit. This person will also be instrumental in communicating best practices to other facilities that were identified in an audit.
- *Facility Manager*: The person responsible for facility operations will be able to directly influence the corrective action needed to address audit findings. The Facility Manager will be able to hold management within the facility accountable for correcting deficiencies and will be able to allocate funding and resources to correct deficiencies.
- *Facility Safety Manager*: In the event of a safety audit, the Safety Manager will be able to manage the ongoing process of addressing audit findings. The Safety Manager will be able to make program and safety training corrections as needed and will work with the Maintenance Manager to correct physical condition issues.
- *Facility Security Manager*: In the event of a security audit, the Security Manager will be able to manage the ongoing process of addressing audit findings. The Security Manager will be able to make program and security training corrections as needed and will work with the Maintenance Manager to correct physical condition issues.
- *Chief*: Fire service audit distribution lists will be based on the unique organizational structure of the district. Appropriate levels of leadership will need to be identified to ensure that accountability is maintained at city and station levels. Once identified, these officers can hold the appropriate individuals responsible for correcting audit deficiencies.

FOLLOW-UP

The final step in the audit process will be following up with the facility to ensure that audit deficiencies are being addressed. Though a great deal of effort and attention is directed at the audit while it is occurring, routine production issues may distract facility management from the audit once it is concluded and a score has been assessed. The auditor can provide assistance by helping facility management to maintain a focus on correcting audit deficiencies through a well-designed system of follow-up. The auditor has several tools to use in accomplishing this task, such as:

- *Prioritization of Findings*: Facility Management may need assistance with prioritizing the list of audit findings. The auditor can assist by conducting a risk analysis of the findings to determine which items should be addressed immediately, which ones can be addressed within the next few months, and which ones can be addressed at an even later time.
- *E-mail*: The auditor can consistently maintain contact with the facility using e-mail. The auditor can request updates on correcting issues, and the facility can use e-mail to ask the auditor questions. E-mail communication can be used on a routine or periodic basis. The auditor can remain in contact

with the location monthly by e-mailing a request for information or updating the location on previous requests for information. E-mail can be used on a periodic basis where the auditor might follow up on the progress of a capital project that is underway, such as the installation of a complex fall protection system.

- *Conferences*: Conference calls or videoconferences can be scheduled to meet with key personnel who are responsible for correcting audit deficiencies. Beyond the use of telephones for such activity, video conferencing technology is available through widely accessible Internet platforms.
- *Periodic Reports*: The audit program may dictate the submission of periodic reports to the auditor on what is being done to correct audit deficiencies. These reports can be submitted via e-mail or created in a web-based format that allows facility management to access a website and enter the appropriate information. The auditor would then review the reports in a timely manner and contact the facility to provide any support that might be needed.
- *Site Visits*: The auditor could return to the site for a follow-up visit. The auditor could walk through the facility to observe progress that has been made in correcting audit deficiencies. The auditor could also review documentation to determine if the facility has made the appropriate changes to address deficiencies found during the audit. Though this option among those available to conduct follow-up might be the most expensive due to the time and travel required, it might also prove to be the most effective given the face-to-face communication and direct accessibility that location management and employees will have to the auditor.

The audit is not complete once the audit report has been distributed. The auditor will need to remain in contact with the facility to ensure corrections are being made. A wide range of tools are available to conduct follow-up. The auditor will need to be aware of these tools and utilize them based on the need and probability of a successful outcome.

CASE STUDIES

FIRE

Tammy has completed the audit for Station 1 and sent a copy to everyone on the distribution list. She proofread the document to ensure that all the information was accurate prior to sending the report. A week later, she receives a call from John, the Training Officer who is responsible for managing everything related to the audit, including correcting deficiencies. The audit has numerous findings, ranging from minor issues to those that will require significant resources to correct. He is unsure of where to start.

SAFETY

Acme Manufacturing has determined that all annual safety audits will be managed as attorney-client privileged. The auditor will be responsible for conducting the audit and completing the report to be submitted directly to the in-house legal counsel. The in-house legal counsel will then distribute the report to a select group of people within the organization. Shelly, the manager for the east region, would like to forward the audit report to all her plants so they can learn from what occurred in the audit and become better prepared for their audits, which are scheduled for later in the year.

SECURITY

Dave is the new Security Manager for Acme Logistics. He is a recent graduate with a bachelor's degree in security management and has only been with Acme for 3 months. Dave had not yet been hired when the annual security audit was conducted 6 months ago. He is appreciative of the effort the auditor has made to follow up through e-mail. However, Dave's lack of experience has caused him difficulty correcting certain deficiencies. He contacts the auditor by phone and requests a site visit by the auditor so he can work with the auditor in person to gain a clear understanding of the issues and potential solutions.

- How would you respond when given the information in these scenarios?
- What options are available to resolve the issues?

EXERCISES

For the following questions, identify a single facility environment in which you would like to situate your responses and answer each question accordingly. Answer each question in the context of a fire, safety, or security audit.

1. Why is it important to complete the audit document as soon as possible following an audit?
2. Would you prefer for audits to be protected from discovery in a court case? Why or why not?
3. Who would you include on an audit report distribution list? Why?
4. Is it important to prioritize audit findings for a facility? Why or why not?
5. What avenues are available for an auditor to follow up with the facility following an audit? Which would be most useful for you? Who would the auditor's point of contact be to follow up with?

Part V

Management System
Audit Development

17 Creating an Audit Document

An audit document will serve as the actual tool that will be utilized to conduct the audit. There are many options available from which to select in relation to the software program or format utilized to construct the audit document. For the purposes of this discussion, Microsoft Word and Microsoft Excel will be utilized to walk through the process of creating audit documents. This is due to the accessibility of these tools as well as understanding the transferable elements that can be used in developing audit documents using other platforms.

ORGANIZATION

Before beginning the construction of the audit document, it will first be necessary to commit time to establishing a proper organization for the product that is intended to be created. In one situation, Microsoft Excel was used to create an audit protocol. The audit was very complex and detailed, but it failed miserably in organization. The audit required the navigation of multiple tabs and multiple areas within a tab to access all the information on a single topic. This poor organization of the audit document was frustrating for the facilities being audited because it was very difficult to identify all the requirements of the audit and to understand what was expected. The audit should be well organized and flow in a logical sequence that allows both the auditor and those being audited to have a clear understanding of the process.

Audit document organization can be accomplished using several strategies. One strategy is to utilize a chronological flow of audit activity. The audit can be visualized as a series of steps that occur in a natural sequence, which can then be transferred to written form in the audit document. The chronology of events can be recorded in the audit document, which naturally progresses from one event to the next and works through the process until all issues covered within the scope of the audit have been addressed.

A second organizational strategy is to utilize a topical grouping of areas to be audited. For example, when creating a safety audit, it may be necessary to cover a large number of topics, including such things as:

- Emergency Action Plans
- Fire Safety
- Lockout/Tagout
- Confined Space
- Powered Motor Vehicles
- Fall Protection
- Hazard Communication

DOI: 10.1201/9781003371465-22

- Hot Work
- Bloodborne Pathogens
- Ergonomics
- Safety Leadership in the Organization
- Incident Investigations
- Drug and Alcohol Testing
- First Responder Team
- Safety Committees

Each of these topical areas can be identified as independent sections within the audit, and all compliance issues for a given topic will be listed in that section. This will include all aspects of the audit:

- Documentation Review (examination of written programs and all supporting documentation that indicates implementation of programs)
- Facility Inspection (examination of the physical manifestations of the programs being implemented in the work environment)
- Employee Interviews (interviewing employees to confirm their knowledge of the content of programs)

Though this strategy may be ideal for communicating an understanding of audit requirements, it may prove difficult when executing the audit. For example, documentation review will typically occur among all programs at a certain stage of the audit. During this time, the auditor will review written programs, training records, and records utilized to implement a program, such as a forklift pre-use inspection form. This strategy will require the auditor to complete only the documentation review questions in each topical area, leaving each section incomplete until the facility inspection and employee interviews can be completed.

A third organization strategy is a blend of the previous two strategies in that it establishes core components of the audit, which are then subdivided by topical areas. These core components can be organized in chronological order of occurrence during the audit. The audit should include the following core components:

- Documentation Review
- Facility Inspection
- Employee Interviews

These three core components, conducted in the order in which they appear in this list, will allow each component to build on the next so that the auditor can gain a holistic perspective of what is occurring at the facility. Performing the documentation review first will allow the auditor to establish a baseline for what activity is occurring at the facility. Written programs will indicate procedures, tools, and training that the facility states are in place to manage a given issue. Once the auditor understands the content of the documentation that is in place, the facility inspection can take place, which allows the opportunity to verify if what is written in the documentation is occurring. If a written Powered Motor Vehicle

Program and training material indicate that forklift operators will stop and blow their horn when coming to the end of a high storage aisle of racking, then the auditor can use the facility inspection as an opportunity to verify if this requirement is being executed in daily production. The auditor can then utilize the knowledge gained during the documentation review and plant inspection to effectively interview employees. Knowledge of employees can be verified, and issues can be more deeply explored based on what has been discovered during the documentation review and the facility inspection.

The three core components of documentation review, facility inspection, and employee interviews can be subdivided to address each topical area covered within the scope of the audit. For example, if an auditor is reviewing a fall protection program, the following would be investigated in the documentation review:

- Written program
- Employee training records
- Contractor orientation documentation
- Equipment inspection and maintenance documentation

The plant inspection would then occur, which would include a review of the following information:

- Observing the use of fall protection equipment
- Inspecting anchorage points that are being used
- Inspecting for proper safety signs and tags
- Inspecting for appropriate handrails and other engineered protection against employee falls
- Inspecting the location and condition of fall protection equipment storage areas

Employees can then be interviewed to ask questions regarding such things as:

- Areas of concern found during the facility inspection
- Confirming the content and nature of training that was documented
- Verification of employee knowledge of procedures, such as emergency response to a co-worker who has fallen and is suspended by a harness and lanyard

Though this audit organization strategy requires a matrix approach to creating and carrying out the audit, the auditor and those being audited should be able to progress smoothly through the audit by navigating each topic within a major component of the audit and then advancing to the next component.

Other strategies may be identified and developed based on the organization being audited, the scope of the audit, and the unique issues that are involved. The goal will be to set the process up for success by organizing the audit in a way that has meaning for the auditors and those being audited. The document should be easy to navigate and explain exactly what is expected.

WRITING AUDIT QUESTIONS

Once the organization of the audit has been established, the remaining competency is to effectively write audit questions. The word "question" may be misleading. Each line in an audit can be phrased as a question or a statement. For example, both these examples would be considered acceptable:

- Are emergency exits clear of obstructions?
- Emergency exits are clear of obstructions.

A "yes" response to the first example that is phrased as an actual question answers the question by indicating that exits are clear of obstruction. A "yes" response to the second example that is phrased in the form of a statement indicates that the statement is true and exits are clear of obstruction.

The first objective for writing audit questions is that the response to every question must be uniform, whether it is 10 questions in a basic audit or 500 questions in a complex audit. It is preferable that between "yes" and "no" responses, the answer should be a positive response of "yes." This would indicate that the item addressed in the question is being properly managed. Following is an example of two questions/statements that could cause confusion:

Question 1
- Are emergency exits blocked?
- Emergency exits are blocked.

Question 2
- Are exit pathways clear?
- Exit pathways are clear.

The issue with these two questions is that safe responses to them are different. The safe response to the first would be "no," emergency exits are not blocked. The safe response to the second would be "yes," exit pathways are clear. A proper response to the first set is "no" and a proper response to the second set is "yes." This can create a great deal of confusion when attempting to score the section of the audit in which these questions reside as well as the audit as a whole. Ensuring each proper response is uniformly "yes" will allow for an error-free environment where all "yes" responses can be considered as favorable and all "no" responses can be considered as unfavorable.

The second objective for writing audit questions is that each question must refer to only one issue. Including two topics within one question can create difficulty in properly assessing the item. For example, the following includes two topics within one question/statement:

- Are fall protection harnesses inspected and properly worn?
- Fall protection harnesses have been inspected and are properly worn.

The word "and" in this example can create a problem in assessing the question. The first portion of the question might be "yes," meaning fall protection harnesses have been inspected, but the second portion of the question might be "no," meaning fall protection harnesses are not being worn properly. Due to this conflict, it is difficult to know whether the question should receive a "yes" or "no" response. This issue can be avoided by separating the issues into two distinct questions/statements:

Question 1
- Are fall protection harnesses inspected?
- Fall protection harnesses are inspected.

Question 2
- Are fall protection harnesses properly worn?
- Fall protection harnesses are properly worn.

The third and final objective for writing audit questions is that they must be written in sufficient volume to adequately address the topic. There is no established number of questions that should be provided to address a given topic. However, care should be taken to balance covering all the issues of a given topic with the compulsion to cover every detail. For example, one question/statement can be utilized to cover training:

- Has employee training for Hazard Communication been conducted?
- Employee training for Hazard Communication has been conducted.

The expertise of the auditor will be needed to evaluate if the training covered all the appropriate requirements of OSHA's Hazard Communication regulation, such as the content of the written program and location of Safety Data Sheets. Rather than needing individual questions for each item delineated in the training requirements, the auditor can sufficiently respond to the question with a "yes" or "no" and then make comments to qualify issues as needed.

Effectively summarizing general sections of policies or regulations to be audited into single questions will help streamline the audit. Rather than having numerous questions for a single item, one question can be well designed to address the key issue at hand, and then comments can be used to provide detailed information to more completely define what has occurred.

BASIC AUDITS

Operational audits that are used on a daily, weekly, or monthly basis will be basic in form compared to a complex management system audit that might occur on an annual basis. Basic audits might include such things as:

- Behavioral observation audits
- Housekeeping audits
- Equipment condition and functionality audits

These types of audits are conducted on a frequent basis and are designed to ensure the ongoing functionality of the item being audited. Microsoft Word can be utilized to create a basic audit document for these types of audits. Creating a housekeeping audit will be the specific example for the purpose of this discussion.

When opening Microsoft Word, a blank document will naturally appear on the screen. Three primary steps will be utilized to create the basic audit document:

- Create a title
- Create a table that provides information to be audited
- Create a record of who conducted the audit

CREATE A TITLE

The title of the document can be created in a very basic form, or details can be provided to make the title more visually appealing. A title will be created that provides a degree of visual interest. This will require the use of a text box. To create a text box:

- Click on the "Insert" tab in the top tool bar.
- Click on the "Text Box" icon.
- Click on "Simple Text Box" in the drop-down menu.

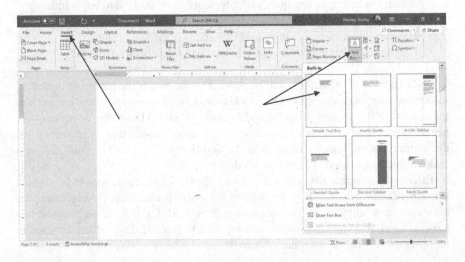

This process will place the text box in the center of the screen. The text in the text box is already highlighted, so the new title of the document can be typed in the text box. In this case, the title will be "Housekeeping Audit." The font and size of the text can be selected and changed:

- Click on the "Home" tab.
- Click on the border of the text box one time so that the dots appear around the edge

- Select the font and size from the respective drop-down menus
- Select a shade to fill the text box by clicking on the "paint can" icon in the tool bar and select a color from the drop-down menu
- Position the text box on the page by clicking on the border of the text box, holding the left button on the mouse down, dragging the text box to the desired location, and releasing the mouse button
- Re-size the text box by clicking on a dot in the center of a border line, holding the button on the mouse down, and dragging the line to the desired size. The box can be resized in the same proportion by performing the same task but by clicking on the dots at the corner of the text box.

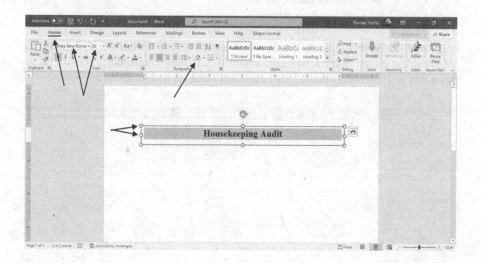

CREATING A TABLE

Utilizing tables in the creation of a basic audit document will help to accomplish two goals. First, tables provide a sense of the visual organization of the document. Second, they provide a form that is electronically functional in that the user can easily click on various sections of the document and enter the appropriate information. A table can be inserted into the audit document by doing the following:

- Place the cursor on the line where the table is to begin
- Click on the "Insert" tab
- Click on the "Table" icon
- Click on the upper right cell and pass the cursor across the grid that is displayed in the drop-down box to select the appropriate number of rows and columns, and then release the click on the box that is on the lowest right side of the grid

In this example 5 columns have been selected that will represent:

- Column 1 –question
- Column 2 – a "yes" response
- Column 3 – a "no" response
- Column 4 – a "not applicable" response
- Colum 5 – space provided for comments to be recorded by the auditor for each question

In this example 8 lines were selected in which to enter:

- Line 1 – headings for each column
- Lines 2-8 – individual questions

Using the grid option limits the number of rows to 8 and the number of columns to 10. Larger tables can be created by clicking on "Insert Table" under the grid in the drop-down box and manually entering the number of columns and rows that are needed.

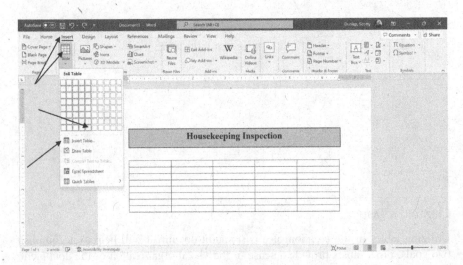

Column titles should be added to the first row of the table. This can be accomplished by:

- Click on the top row of each column and type the word for each of the column titles
- Highlight the row, click on the "Center" icon in the top tool bar, and select the font and size desired
- With the row still highlighted, remove the grid lines from the title row by clicking on the grid line icon and click on "No Border." Click on the grid line icon again and select "Bottom Border" to restore the line that serves to separate the title row from the row that will contain the first question.

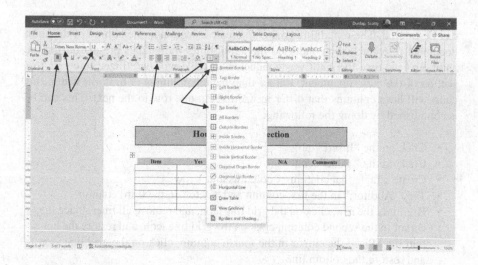

The table that is displayed will contain columns that are uniform in width, so it will be necessary to adjust each column to accommodate the desired width. This can be accomplished by:

- Place the cursor in any of the rows within a column that needs to be adjusted
- Click and hold the mouse button down on the marker in the document ruler for the side of the column that needs to be adjusted. Drag the marker to the desired width and release the mouse button.
- Cosmetically, the "Yes," "No," and "N/A" columns should be uniform in width. Rather than attempting to accomplish this manually, highlight the three columns, press the right mouse button, and select "Distribute Columns Evenly."

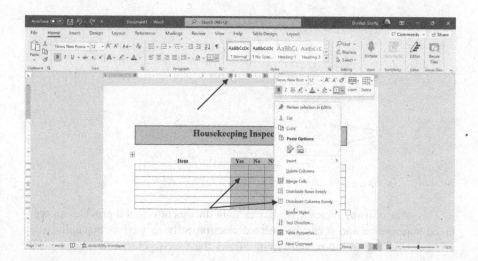

The table to be used for the audit has now been created. Questions for each row can be entered. The remaining goal is to create a section at the bottom of the audit document to record who conducted the audit, other individuals present during the audit, and the date the audit was conducted. A table will be utilized to create this section of the audit document. Rather than having columns of uniform width, this chart will have columns that differ in width from one row to the next. This can be accomplished by doing the following:

- Click on the "Insert" tab
- Click on the "Table" icon
- Create a table that contains 4 columns and 1 row
- Type "Auditor:" in the first column and "Date:" in the third column
- Highlight the row, click on the grid line icon, and remove all lines. Place the cursor in the second column, click on the grid line icon, and restore the bottom line. Place the cursor in the fourth column, click on the grid line icon and restore the bottom line.
- Place the cursor in each column and manually adjust the column width to the desired size.
- Strike the "Enter" key to create a space below the table and create another table that contains 2 columns and 1 row
- Type "Individuals Present:" in the first column
- Highlight the row, click on the grid line icon, and remove all lines. Place the cursor in the second column, click on the grid line icon, and restore the bottom line.
- Place the cursor in each column and manually adjust the column width to the desired size.

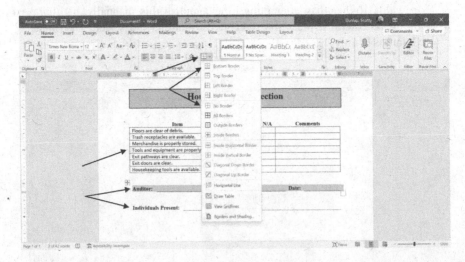

The value of using tables to design an audit document is that it provides an organized appearance and it can be utilized electronically as well as manually when printed. Clicking in a box and typing allows the user to enter information while maintaining the presence of lines within the design of the chart. In sections that identify the auditor, date, and those present, one option is to simply type the words, enter a space and then press the "Shift" and "underline" keys to make the line on which an auditor can enter the information. Though this will be effective when utilizing the audit form to write on as a hard copy, the underline will not remain as designed when attempting to type on it electronically. Utilizing a table will provide clean formats for hard copy and electronic use.

COMPLEX AUDITS

Where basic audits can typically be accomplished by documenting activity on limited sheets of paper, a complex audit is much more intricate. Complex audits might include things such as:

- Annual safety management system audits
- Annual security management system audits
- Annual fire service audits

These types of audits are conducted on an infrequent basis and are designed to establish how the system is performing at the point in time when the audit is conducted. Microsoft Excel can be utilized to create a complex audit document for these types of audits. Creating a safety audit will be the specific example for the purpose of this discussion.

When opening Microsoft Excel, a blank spreadsheet will naturally appear on the screen. The following primary steps will be utilized to create the complex audit document:

- Create a title page
- Insert tabs that represent core sections of the audit
- Create tables that provide information to be audited
- Create formulas that tabulate section scores and total scores that are referenced on the title page
- Create instructive text that assists facility personnel in understanding what needs to be done to receive full credit for a given question

The audit to be created will use the organizational strategy that establishes core sections of the audit with sub-sections that identify topics. The core topics will be:

- Documentation review
- Facility inspection
- Employee interviews

The scoring methodology used in this initial example will utilize the following for each question:

- Yes – 1 point
- No – 0 points

CREATE A TITLE PAGE

Creating the title page will be the first opportunity to exercise organizational skills. When an audit is performed and the audit document is completed, high-ranking executives or officers in the organization may not have time to read through the numerous sections of the audit. The title page is an opportunity to provide a high-level understanding of how the facility performed. High-ranking executives can utilize this portion of the document as a summary to gain an immediate understanding of how the facility performed.

The title page is highly customizable to meet the unique needs of an organization. The following general items should be considered:

- Audit title
- Summary score for each core section of the audit
- Composite score for the audit
- Primary findings
- Primary best practices
- Individuals involved in conducting the audit
- Title the tab

Stephen Covey researched success literature that was written in the first 150 years of our nation's history to determine factors for success that were identified in that period of time compared to the get-rich-quick literature that has come into vogue in our nation's recent history. He published his findings in his pinnacle work, titled *The Seven Habits of Highly Effective People*. One of the seven habits that Covey identified was "Begin with the End in Mind." Beginning with the end in mind requires visualizing the product that should be created. This habit of success can be utilized when constructing the title page because it will serve as an executive summary of how the facility performed on the audit. Thought will have to be given as to how this page can be designed to most effectively communicate the necessary information.

AUDIT TITLE

The title of the audit should be direct, concise, and accurately reflect the scope of the audit. For the example used here, the title will be *Safety and Health Audit*. The word "environment" is often utilized in conjunction with "safety and health." A distinction needs to be made in relation to these words. The words "safety and health" are typically used in reference to issues governed by the Occupational Safety and Health Administration (OSHA). Though OSHA is concerned with the work environment, the word "environment" is typically used in reference to issues governed by the Environmental Protection Agency (EPA), which oversees the natural environment. The use of the word "environment" in the title of the audit is acceptable if EPA issues are covered within the scope of the audit.

Entering the title of the audit into the first tab of the Excel spreadsheet can be accomplished by the following steps:

- Highlight the number of cells in the first row that comprise the width of the title page
- Click on the merge cell icon in the top tool bar and select "Merge Cells"

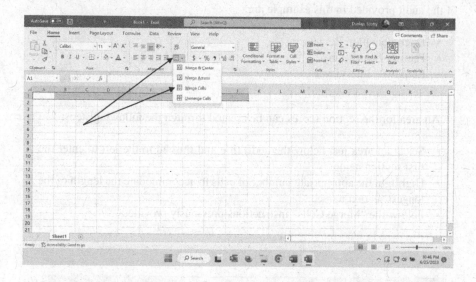

The title will need to be entered into the merged cell and then set in a larger font size and in bold print. This can be accomplished by doing the following:

- Click on the merged cell
- Select the font size that is desired
- Click on the "Bold" icon in the top tool bar
- Type the text "Safety and Health Audit"

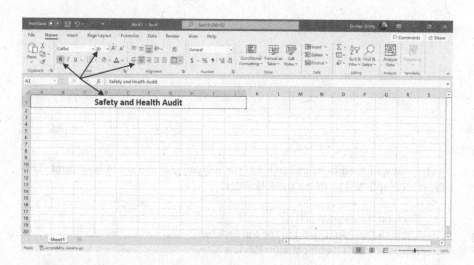

SECTION AND COMPOSITE SCORES

Providing scores for each section of the audit allows the reader to see at a glance how the facility performed in each of the primary sections of the audit. The three sections of the audit provided in this example are:

- Documentation Review
- Facility Inspection
- Employee Interviews

An area for the section scores can be created through the following steps:

- Select an area just below the audit title and flushed to the left to enter the first section name
- Highlight the appropriate number of cells to accommodate the length of the longest section name
- Follow the "Merge Cells" instructions previously provided

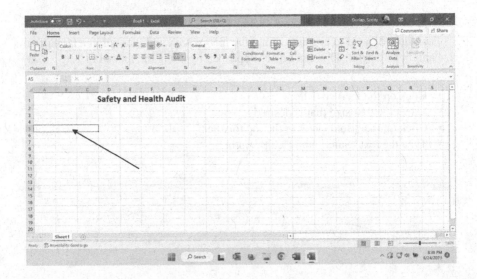

The title will need to be entered into the merged cell and set in bold print. This can be accomplished by doing the following:

- Click on the merged cell
- Click on the "Bold" icon in the top tool bar
- Type the text "Documentation Review"

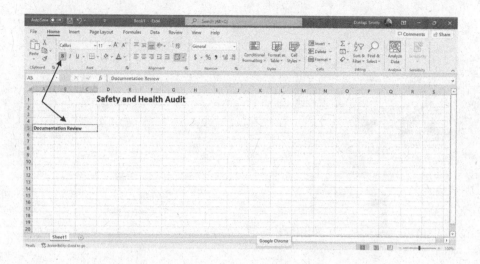

This series of steps must now be repeated to enter headings for "Facility Inspection" and "Employee Interviews." Once these have been entered, the series of steps can be repeated to enter a line for "Composite Score."

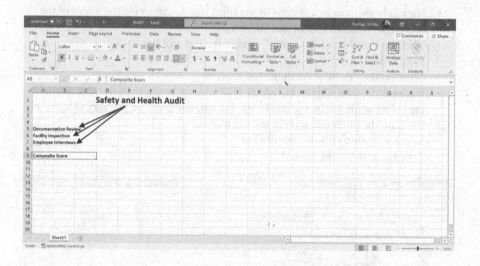

The final step for this section of the title page is to create grid lines for cells that will contain the scores for each line. This can be accomplished through the following steps:

- Highlight the cells that will contain grid lines
- Click on the grid line icon in the top toolbar
- Select the image that depicts grid lines on "All Borders"

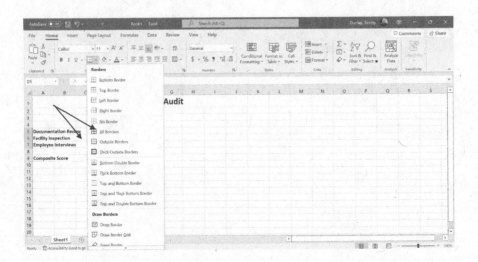

To give the title page more visual interest and a clearer delineation between sections, a shaded bar with a section subtitle can be added. This can be accomplished by performing the following steps, which utilize several steps previously presented as well as one new option:

- Highlight the number of cells in the third row to comprise the width of the title page (equal to that of the title line)
- Click on the "Merge & Center" icon in the top tool bar
- Select "Merge Cells"
- Click on the "Bold" icon in the top tool bar
- Increase the font size to 14, which will make it larger than the font size of the information in the section below but smaller than the primary title
- Click on the paint can icon in the top tool bar
- Select a shade to fill the subtitle bar
- Type the text "Audit Scored"

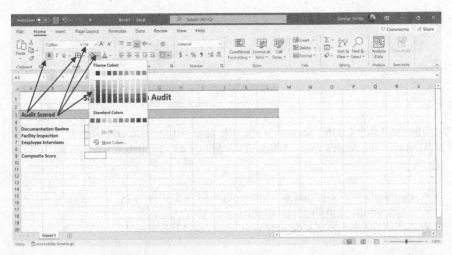

PRIMARY FINDINGS AND BEST PRACTICES

Primary findings from the audit can be added to provide a high-level perspective on issues that surfaced during the audit. This component of the executive summary of information on the title page will alert high-ranking executives or officers to the main issues that need to be addressed at the facility that was audited.

A section for best practices can be added to provide a balanced view of what is occurring at the facility being audited. Rather than the audit being an exercise to identify problems at a given facility, it can also be a tool to identify best practices. This will help to serve two purposes. First, rather than only chastising a facility for things that were found to be wrong, the identification of best practices will help to acknowledge the good work that is occurring at the facility. Second, the communication of best practices to other facilities can help improve organizational performance.

The process of creating sections for this information is similar to that used for creating the scoring sections:

- Select an area just below the audit scoring section to enter the "Primary Findings" subtitle heading bar
- Highlight the number of cells needed to type "Finding One" below the subtitle heading bar and flushed it to the left
- Click on the merge cells icon in the top tool bar
- Select "Merge Cells"
- Click on the bold icon in the top toolbar
- Type "Finding One"
- Merge the remaining cells, starting with the cell immediately to the right of "Finding One" continuing to the right border of the title page
- Repeat the previous steps to create lines for "Finding Two" and "Finding Three"
- Highlight the three merged cell areas to the right of the "Finding" titles and click on the borders icon in the top tool bar to provide borders on all sides

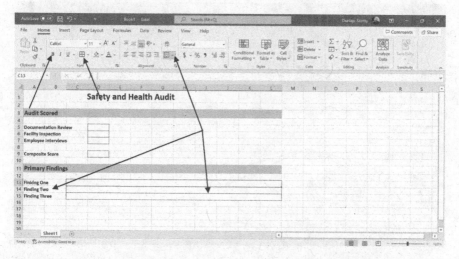

The process can be repeated to enter a section for "Best Practices."

THOSE INVOLVED IN CONDUCTING THE AUDIT

It is important to document who was present during the audit. This can be accomplished on the cover page by creating a table that delineates the auditor(s) and others who were present at the time of the audit. This can be accomplished by following steps like the previous two sections of the title page:

- Select an area just below the "Best Practice" section to enter the "Individuals Present" subtitle heading bar
- Highlight the number of cells needed to type the names of individuals flushed to the left below the subtitle heading and merge the cells using the "Merge & Center" option
- Click on the bold icon in the top toolbar
- Type "Names" as a column title
- "Merge & Center" the number of cells immediately to the right of the "Names" merged cells that will be needed to accommodate the role of each person that was present
- Click on the bold icon in the top toolbar
- Type "Role" as a column title
- Merge each group of cells in the following 6 rows under each column heading by highlighting all of the cells in the column and the number of rows desired, and then click on "Merge Across" in the merge cells dropdown menu. This will provide the ability to merge cells in multiple rows in one step, rather than merging each row individually.
- Highlight each of the empty merged cell groups and click on the border icon in the top tool bar to provide borders on all sides

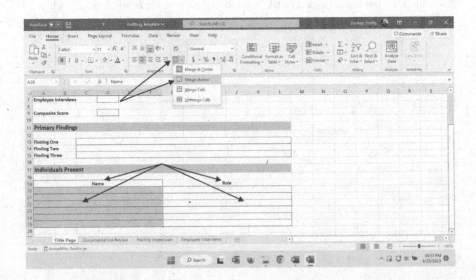

TITLE THE TAB

The final step in creating the title page is to provide a title on the spreadsheet tab. This can be accomplished by the following steps:

- Place the cursor over "Sheet 1" at the bottom of the screen and depress the right click button on the mouse
- Click on "Rename"
- "Sheet 1" will now appear highlighted. Type "Title Page."

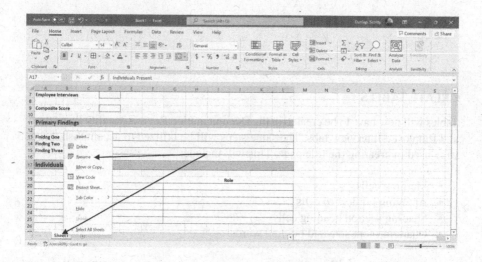

The title page has now been created. The steps utilized in creating the title page will be utilized in creating the remainder of the audit, with the additional skill of creating mathematical formulas and referencing formulas in various parts of the spreadsheet. The steps in this section will not be repeated, so they may need to be referenced as the remainder of the process of creating an audit document is navigated.

INSERT TABS

This sample audit will use the organizational strategy of identifying the three primary categories of Documentation Review, Facility Inspection, and Employee Interviews. Including the title page, there will need to be four tabs to accommodate a tab for each item. The spreadsheet naturally opens with three tabs at the bottom of the screen. The fourth tab can be added by simply clicking on the icon that appears to the right of the three tabs. Each tab can then be renamed to identify its topical area of the audit.

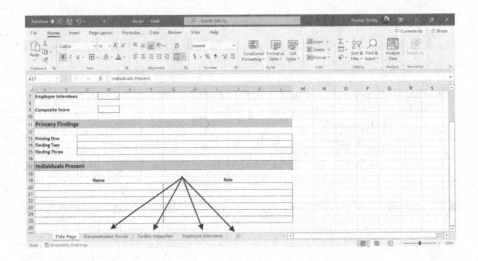

CREATE TABLES

Tables will now have to be created in the Documentation Review, Facility Inspection, and Employee Interview tabs. The same steps will be utilized to accomplish this as were used in creating the title page. This will include:

- Merging cells
- Applying borders to cells
- Shading section heading cells
- Changing font size and bolding words

A new skill that will be applied to this section is the use of mathematical formulas to calculate scores among subsections of the audit. These scores will then be used to calculate the primary section scores and the audit composite score on the title page.

The first step will be to create primary and subsection headings on the Documentation Review tab that represent each topic to be covered within the scope of the audit. This can be done by doing the following:

- Merge the same cells across the spreadsheet as was done to form the subsection headings on the title page. The width of this area will need to be wider than that of the title page to accommodate the volume of columns and information that will be entered into this tab of the audit document. One function that will save time is to highlight several rows and columns that need to be merged to create a number of lines and select "Merge Across." This will merge all cells horizontally but allow each row to remain intact.
- Apply a border around the merged area
- Fill the title line area with the same shade as was used on the title page
- Adjust the title line font size to 14 point
- Enter the title of the topic

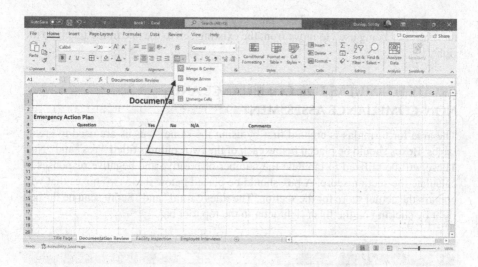

With the initial organization of the section complete, borders can be placed in the table and questions for the documentation review for that subsection can be entered. When the audit is conducted, the auditor will type comments into the far-right column to clarify issues that may not be simply addressed through a "yes" or "no" response in the appropriate column. It will be necessary to wrap the text in the "Comment" cells so that the text will perform an automatic return when the line reaches the right side of the given cell. This can be accomplished by highlighting the "Comment" cells and clicking on the "Wrap Text" icon in the top tool bar "Format" icon.

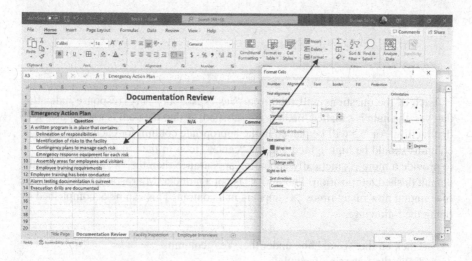

The first question opens with a primary statement, but under it are six questions that delineate specific items. The indentation for each of those questions is accomplished by clicking on the cell and striking the space bar a uniform number of times to establish an even margin. The "tab" key cannot be used for this purpose due to the "tab" key being used to move from the existing cell to the one immediately to the right.

CREATE FORMULAS FOR COMPLIANCE AND NON-COMPLIANCE ASSESSMENT

The area for formulas is created first by using the same steps as are used to create tables. Boxes should be placed below each of the two columns titled "Yes" and "No." These will be utilized to tabulate the number of positive and negative responses to calculate the section score. A box should be placed below these two boxes that will contain the actual score for the section. The titles "Total" and "Score" can be flushed right by clicking on the flush right icon in the top tool bar.

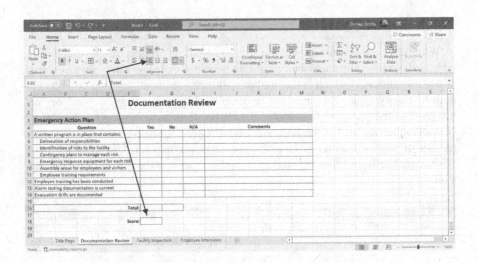

Each of the questions will require a "Yes," "No," or "N/A" response indicating that yes, the item is satisfied, no, it is not satisfied, or that the item is not applicable to that facility. An "x" will be placed in the appropriate column for each question. The columns should be highlighted and the "center" icon in the top tool bar clicked on to provide a more cosmetically appealing appearance than the "x" appearing in the default flushed left position. Though "x" is not a number, Microsoft Excel is capable of counting how many times "x" appears in a column. This can be accomplished by doing the following:

- Click on the "Total" box under the "Yes" column
- Enter the following formula:

$$= \text{COUNTIF}(\text{F6}:\text{F14}, \text{``x''})$$

There are four things to understand about this formula:

- The equal sign establishes what is to come afterward as a mathematical formula that the cell will use to calculate what is needed
- COUNTIF is a command that indicates the formula will search for a symbol and count the number of times it appears
- F6:F14 is the range of cells covered by the formula. In this case, it is the "Yes" column for each of the questions contained in the Emergency Action Plan subsection of the audit.
- The symbol that the formula searches for to count is represented by "x." It is important that the quotation marks appear around the symbol.

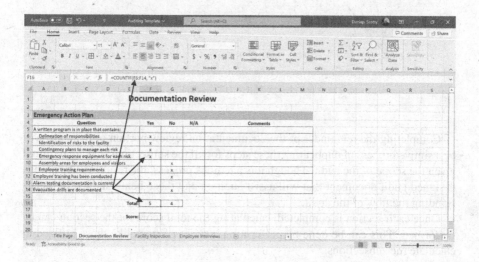

The five total "yes" responses added to the four total "no" responses equal the total number of questions that apply to the facility for this subsection of the audit. The score for this subsection is calculated by dividing the number of "yes" responses (5) by the number of total applicable questions (9), which is the total "yes" responses added to the total "no" responses. This can be represented in the score box by clicking on the score box and then clicking on the percent icon in the top tool bar. The formula for the "Score" cell is as follows:

- =F16/(F16+G16)

There are four things to understand about this formula:

- The equal sign establishes what is to come afterward as a mathematical formula that the cell will use to calculate what is needed
- F16 represents the cell containing the total number of "yes" responses
- /is the mathematical symbol for division

- (F16+G16) performs addition among the two cells containing the total number of "yes" responses and the total number of "no" responses

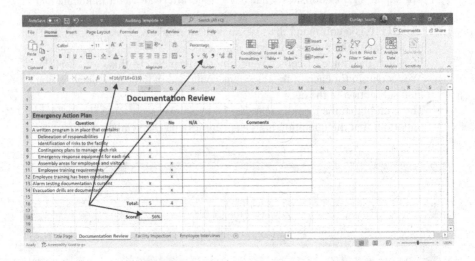

Completing the remainder of the documentation review portion of the audit will be a simple process of duplicating these steps for all the subsections that are required given the scope of the audit. Though it is a tedious process, the result will be a well-designed audit document that will assist auditors and those being audited in understanding the flow of the audit.

Once each section is completed, calculating the total score for the section can be performed. This can be done by performing a series of steps like those utilized to calculate the subsections:

- Create two lines below the last subsection of the audit that are titled "Grand Total" and "Documentation Review Score." They should be in the same format as the "Total" and "Score" lines and boxes for each of the subsections.
- Click on the "Yes" column box next to "Grand Total." Click on the "AutoSum" icon in the top tool bar. Click on the first "Yes" subsection total." While holding down the "Ctrl" key, click on the remaining "Yes" subsection totals. This will add all the "Yes" subsection totals into one grand total for all "Yes" responses. Strike the "Enter" key after all "Yes" subsection response totals have been selected. Click on the "No" column box next to "Grand Total" and repeat the process by clicking on the "AutoSum" icon and then all "No" subsection totals, and then striking the "Enter" key.
- Click in the box next to "Documentation Review Score" enter a formula like the subsection scoring formula that divides the Grand Total "yes" responses by the Grand Total "yes" responses added to the Grand Total "no" responses. Click on the percentage icon in the top toolbar to present the calculation as a percentage score.

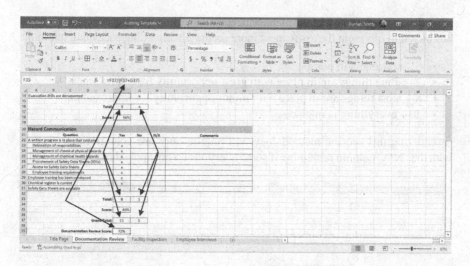

The total score can now be transferred to the title page. This can be accomplished through a series of simple steps:

- Click on the box next to "Documentation Review" on the cover page
- Type the equal sign to set the cell up to calculate a formula
- Click on the "Documentation Review Score" percentage box in the "Documentation Review" tab
- Strike the "Enter" key

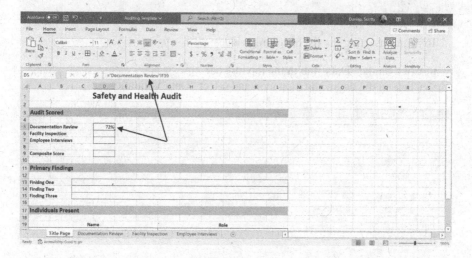

CREATE FORMULAS FOR DETAILED SCORING

Detailed scoring provides a more accurate representation of what is occurring at a facility within the management system. Rather than compliance (yes) and non-compliance (no) being the only two options, quantitative measures can be utilized to provide a more detailed assessment of the management system. The following are scores that can be entered to represent the assessment more accurately regarding each audit question:

- 0 – nothing is in place
- 2 – some things are in place with great opportunity for improvement
- 8 – most things are in place with minor opportunity for improvement
- 10 – everything is in place

Utilizing this scoring methodology will require only one column titled "Score" with the "N/A" column remaining if some questions do not apply during a given audit.

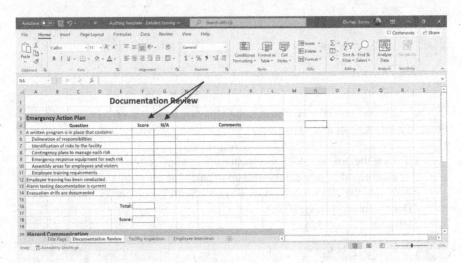

To begin inserting the list of scoring values:

- Click on the "Data" tab in the top tool bar
- Click on the data validation icon and select "Data Validation"

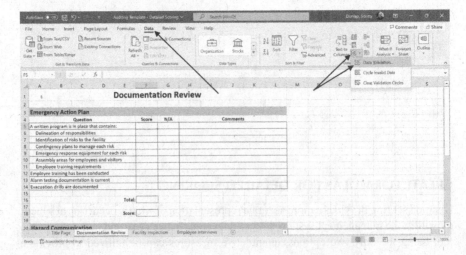

In the Data Validation window that appears:

- Click on the "Settings" tab
- Click on the arrow down icon to display a dropdown menu
- Click on "List"

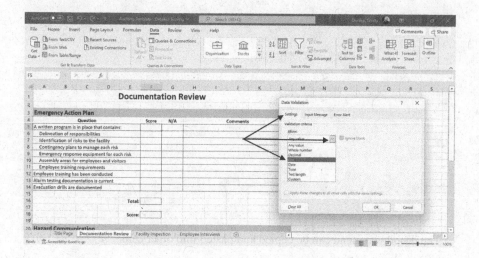

On the "Source" line, type each score value with a comma and space following each value:

- 0, 2, 8, 10
- Click "OK"

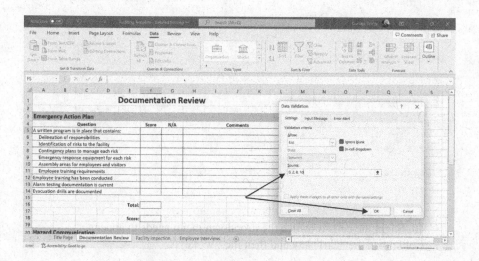

Clicking on the cell will cause a dropdown arrow to appear to the right of the cell. Clicking on the arrow will present a dropdown menu with the potential scoring options that can be clicked on based on the assessment of the question.

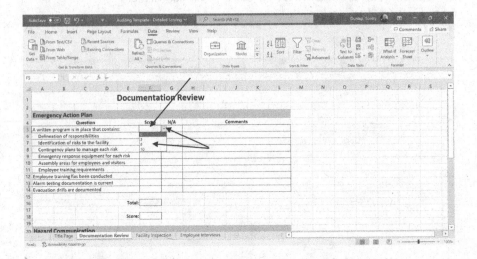

The dropdown menu can be copied and pasted into other cells:

- Right-click on the cell
- Select "Copy"
- Or hold down the "Ctrl" key and then the "C" key

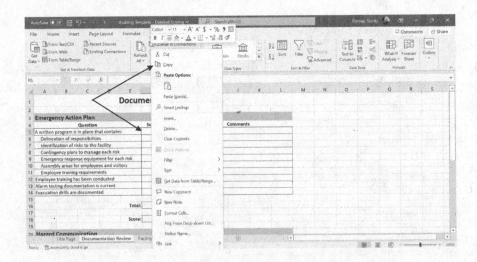

- Highlight the cells in which the dropdown menu needs to appear
- Hold down the "Ctrl" key and then the "V" key

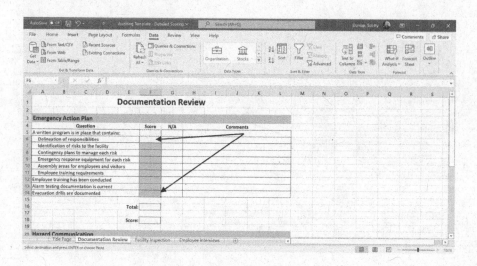

Inserting the formula to tabulate the section scores will include the following steps:

- Click on the cell in which the "Total" score will appear
- Click on the Greek letter that resembles an "E" in the top tool bar
- Click on "Sum"

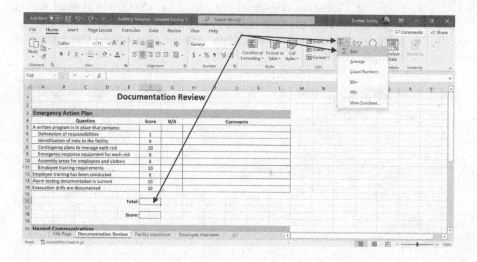

- Highlight the cells that represent scores for each question
- Hit the "Enter" key

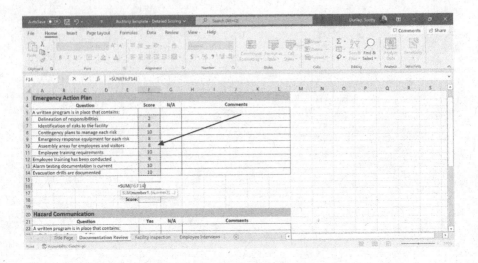

To enter the formula for the "Score" cell:

- Click on the cell in which the score should appear
- Type the formula as the equal sign (=) followed by:
    - F16 – the cell that presents the total points earned
    - /– the symbol that represents the mathematical division function
    - (9*10) – creates a formula for the number of applicable questions in the section (9) multiplied (*) by the maximum number of points possible for each question (10)
    - Note – if a question is deemed not applicable (N/A), it will need to be factored out of the total points possible, causing the divisor in the formula to be altered to be the total number of questions that apply multiplied by 10

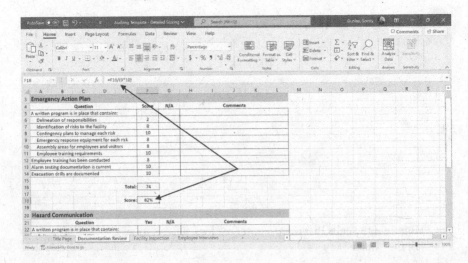

INSERTING INSTRUCTIVE TEXT

Though a specific audit question may appear to be clear in its intent, supplying instructive text for audit questions can assist both auditors and those being audited to understand what must be in place to receive a "yes" response or 10 points, depending on the scoring methodology. This can be done by following a few simple steps:

- Click on the cell that contains the question for which instructive comments need to be added
- Depress the right button on the mouse, and a list will appear. Select "New Note."
- A text box will appear in which wording can be typed that provides compliance information for the question
- Resize the box by clicking on and dragging the corners and/or sides of the text box so that all the text appears

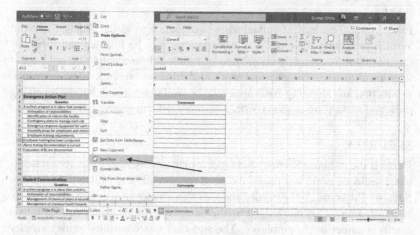

Once the text box has been completed, a small triangle will appear in the upper right corner of the cell, alerting the user that instructive information has been entered for that question. Simply passing the cursor on top of the cell will cause the textbox to appear.

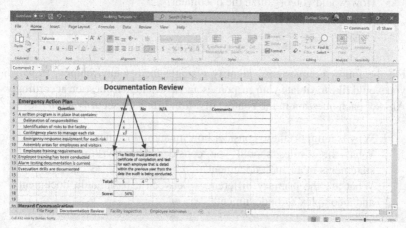

Each of the steps utilized to create the information contained in the "Documentation Review" can be utilized to create the information for the "Facility Inspection" and "Employee Interview" tabs.

CASE STUDIES

FIRE

Jill is a new officer in her department. She has worked in the fire service throughout her career and is concerned for both the public and her fellow fire-fighters. Though her department is engaged a great deal in community education and awareness, she believes there is an opportunity to address the risk that exists within the fire station.

SAFETY

Eddy joined Engitech Manufacturing two years ago as the Safety Manager. Written programs and employee training appear to be in place and functioning, but there is currently no method of thoroughly evaluating them to see if they are effective. The physical work environment presents a few concerns. Production and shipping areas appear to be well maintained, but the receiving dock and maintenance shop routinely display risks to employee injury.

SECURITY

Debbi is the in-house Security Manager for Distro Distribution. A recent workplace violence event has raised the need to more closely evaluate the facility Security Program. A disgruntled spouse of an employee accessed the property because she was acknowledged by the Security Officer at the front gate as a family member of the employee. The situation was controlled prior to injury occurring.

- How might a complex audit document be utilized in these situations?
- How would you organize the document?

EXERCISES

For the following questions, identify a single facility environment in which you would like to situate your responses and answer each question accordingly. Answer each question in the context of a fire, safety, or security audit.

1. Why is it important to consider how an audit document is organized?
2. What three rules should you consider when writing questions/ statements in an audit document?
3. What are the primary differences between a basic audit document and a complex audit document?

4. How do you insert a table into a document that is used to conduct a basic audit?
5. Why should both Primary Findings and Best Practices be included on the title page of the audit document?
6. How do you insert a new tab in a complex audit document to accommodate the number of main sections in your audit document?
7. What formula is used to count the number of times an "x" appears in each section of the audit document?
8. What formula is used to calculate the score for a given section in your audit document? How do you determine the number of points possible if some questions in the section are not applicable to the facility?
9. How can you reference a calculation from another tab in the audit document on the title page?
10. How can you insert instructive text that assists both the auditor and those being audited to understand what is necessary to satisfy the requirements of a given audit question?

18 Audit Scoring

The issue of scoring audits can create opposing opinions within an organization. Some believe that audits should not be scored, while others believe that audits should be scored. Those who believe that audits should not be scored argue that ranking may prove harmful and that a simple list of findings should be generated that the facility would need to address. Those who believe audits should be scored might say the score is a critical component of accountability in that it indicates how well or poorly the facility is performing, and the score provides a metric to measure performance as audits are conducted over time. The answer to this question is not simple. There are variables that should be considered when determining if an audit should or should not be scored.

ORGANIZATIONAL CULTURE

Organizational culture must be understood to appreciate the subtle details that could impact how a score, or lack of a score, would be perceived by those in leadership. In research that was conducted by Fred Manuele on the loss of the space shuttle Columbia, it was found that it is helpful to understand the power of organizational culture and its influence on workplace safety. The loss of the space shuttle Columbia was evaluated from the perspective of dissecting the report issued by the Columbia Accident Investigation Board. The report was a comprehensive explanation of the findings of issues with the shuttle and within NASA that eventually led to the loss of the vehicle. Though the study was published specifically in an occupational safety journal, there are broad applications to leadership in any organization.

One leadership issue that led to the loss was poor organizational communication. Individuals at lower levels in the organization with critical information faced great difficulty in getting it to the decision-makers. Leadership within NASA also transitioned from a philosophy of the mission needing to be proven fully safe prior to moving forward to a philosophy of needing to prove that there is a risk. This was a critical shift in the organizational culture within NASA.

Understanding such cultural issues will assist in the determination of how a score can be implemented within an organization. The fire, safety, or security culture within an organization can have an impact on how well the score from an audit will be perceived.

ACCOUNTABILITY

A goal of the auditing process is to hold facility personnel accountable for performance related to activities within the scope of the audit. The organization will have to evaluate the proper avenue to ensure that accountability is assigned to individuals and is then properly managed. Though a score is one way to achieve accountability,

DOI: 10.1201/9781003371465-23

a list of findings may also achieve accountability when coupled with a performance evaluation process.

A performance evaluation process is a formal tool utilized to determine how well an individual is performing regarding established goals that the individual must achieve and the strategy by which they will be achieved. This process typically begins at the start of an organization's fiscal year, with meetings occurring between the supervisor and subordinate at the following intervals:

- *Initial*: A meeting is held at the beginning of the fiscal year to establish expectations for performance and how those expectations might be achieved.
- *Mid-Year*: A meeting is held at the mid-year point to identify progress that has been made to-date and what activity must be accomplished in the remaining half of the fiscal year.
- *Year-End*: A meeting is held at the end of the fiscal year to determine if performance expectations have been met and what opportunities for improvement might exist.

If an audit score is not utilized, a list of findings for which the individual has responsibility can be utilized to determine if there has been success in achieving performance expectations. Care must be taken to assign accountability over which the individual has complete control. For example, if the audit reveals that the injury rate in a given department has increased from the previous year, it may not be appropriate to hold the department supervisor accountable for the rise in the injury rate. However, the supervisor can be held accountable for activities within his/her control that could affect the injury rate, such as:

- Defined behaviors that facilitate being a positive role model
- Discussing safety topics frequently in pre-shift meetings
- Engaging in incident investigations and supporting corrective measures that will prevent the incident from recurring
- Being involved in the facility safety committee

These behaviors are those over which supervisors have direct control. Supervisors can engage in these activities, which will in turn affect the injury rates within their departments and the facility as a whole.

A score can also be utilized to directly hold a supervisor accountable for performance within the scope of the audit. Due to audits evaluating the system that is in place, items that are evaluated within the scope of the audit are those things over which a supervisor has direct control. This might include:

- *Employee Training*: The supervisor may be responsible for conducting employee training or ensuring that employees can be released from normal work activity to attend training. Reviewing safety topics in pre-shift meetings could also be viewed as a form of training.

- *Program Implementation*: The supervisor has control over holding employees accountable for material on which they have been trained. The supervisor can establish a culture of compliance by executing corrective action in keeping with human resources policies when an employee has violated a fire, safety, or security policy. The supervisor can also encourage employees who are observed to be demonstrating behavior that is in compliance with established programs.
- *Facility Inspection*: The supervisor has control over the physical condition of the department. Housekeeping responsibilities can be assigned to employees and followed up on by the supervisor to ensure a safe and orderly workplace is maintained.

These and other responsibilities of the supervisor can be addressed through accountability for an audit score. Care should be taken to ensure that supervisors are held accountable for those things within the audit score over which they have direct influence.

AUDIT SCORING

If the decision is made to score the audit, it will be necessary to determine the strategy that will be utilized. There are two primary methodologies that can be used. One employs an "all or nothing" methodology where the audit question is either satisfied or not satisfied. The second methodology employs a graded scale that takes into consideration varying levels of compliance.

COMPLIANCE OR NON-COMPLIANCE

The compliance/non-compliance methodology utilizes a basic approach to assessing compliance throughout the audit. This will typically be accomplished by providing points based on the following two options:

- Yes (in compliance) – 1 point
- No (not in compliance) – 0 points

A given audit question might focus on fire extinguishers being inspected on a monthly basis. If any fire extinguishers are found to be in violation, that question would receive a "no" response. If all fire extinguishers are found to be inspected, then that question would receive a "yes" response.

"No" responses result in the assessment of no points. This indicates that the facility has a deficiency, which results in no credit being given within the audit scoring methodology. A "yes" response results in an assessment of 1 point, which indicates the facility receives credit for work done in relation to that question.

SCALED SCORING

The second approach to establishing a scoring methodology employs the use of a scale that can provide a spectrum of points based on the degree to which a facility complies with a given question. This strategy may help to communicate more clearly what is actually occurring compared to an "all or nothing" approach. For example, when evaluating fire extinguisher compliance, it may be found that among the 200 fire extinguishers in a facility, 20 have not been inspected on a monthly basis. The "all or nothing" approach would award zero points by assessing a "no" response to that question due to the 20 deficiencies that were found. Unfortunately, this scoring option might not clearly communicate what is occurring due to the statistics indicating that the facility is actually at 90% compliance on fire extinguisher inspections.

The scaled scoring approach will be accomplished by providing points based on the following:

- *10 points*: The audit question is completely satisfied.
- *8 points*: Compliance with the audit question is primarily achieved, but there were a few issues that were identified.
- *2 points*: Compliance with the audit question was somewhat achieved, but there were a large number of issues that were identified.
- *0 points*: Compliance with the audit question was not achieved.

It is important to notice that on this scale, there is no middle option of 5 points from which the auditor can select. The absence of a 5-point option is by design because it forces the auditor to rate the question predominantly in one direction or the other. The middle ground option might present the auditor with too much temptation to rate a questionable area in the middle. A 0-point, 2-point, 8-point, and 10-point scale causes the auditor to select an option that is toward compliance or toward non-compliance. The 0-point and 10-point options are clearer for the auditor in that either compliance is not achieved, or it is completely achieved whereas the 2-point and 8-point options allow the auditor the freedom to determine if the issues identified weight the response toward non-compliance (2 points) or compliance (8 points). Using the example of the fire extinguisher inspections, assessing the facility at 8 points for primarily compliance with a few issues identified might more accurately reflect the situation when compared to assessing 0 points and a "no" response within the "all or nothing" strategy.

NOT APPLICABLE

Questions that are not applicable to a facility will have to be carefully managed when calculating a score for an audit section and the composite score for the audit as a whole. Questions that are not applicable will need to be factored out of the scoring calculations. For example, a given section of the audit might contain 10 questions. If two questions are not applicable to the facility, that leaves eight questions for which the facility can earn points toward the score. The two questions that are not applicable will need to be factored out of the calculation to arrive at an accurate score. Consider the difference as follows:

	Correct	Incorrect
Yes Responses	6	6
No Responses	2	2
N/A Responses	2	2
Calculation	6/8	6/10
Score	75%	60%

In the "Correct" column, the number of "yes" responses is divided by the sum of the number of "yes" responses and "no" responses. This sum reflects the number of questions that are possible for which the facility can earn points (8). The "Incorrect" column divides the number of "yes" responses by the total number of questions in the section (10). The score of 60% is skewed due to including questions that were not applicable in the calculation, thus penalizing the facility for questions that do not apply to the work environment.

AUDITOR CALIBRATION

Whether an "all or nothing" or scaled scoring methodology is utilized, auditors will need to be calibrated on the proper use of the methodology. This can be accomplished in the auditor training process. Auditors can be trained in the mechanical functioning of the audit scoring process in a classroom setting and then witness its application by being involved in a pilot audit. Judgment will need to be utilized in determining how to apply the scoring methodology in various circumstances. Exercising judgment uniformly among auditors can be achieved through calibration in the classroom and practical skills exercises. Theoretically, a given facility should receive the same score if audited by multiple auditors.

CASE STUDIES

FIRE

An auditing system has been in place in the city fire department for five years, and improvements are being made, but not at the pace that Lou believes could occur. As the Captain at his station, he believes his team should receive recognition for the work that has been put into correcting issues identified in the audit. At this time, the annual audit is not scored, and Lou believes that adding a score will help to delineate the levels of performance of the stations in the city.

SAFETY

Mary has been working diligently to improve the level of safety performance at her facility by responding to audit reports that have been generated through the annual safety audit. She is disappointed that her effort has resulted in only meager increases in performance from one year to the next. A problem that she sees is that she can only do so much to impact the audit score. Department managers are not engaged in the process, but she recognizes their help is instrumental in experiencing significant improvement.

SECURITY

New Wave Software is a progressive organization where employees are encouraged to express their individuality and work in a relaxed atmosphere with few rules. Clark is the Security Manager, and he has noticed an increasing risk of the loss of the products that New Wave has created in recent months. He has been told by the IT Manager that there has been a slight increase in attempts to infiltrate their system. Previous audits have presented a list of findings of which department managers have been made aware, but Clark has not seen an increase in diligence on their part to guard against security threats.

- What organizational culture issues might need to be considered?
- What issues related to the approach to scoring the audit should be considered?

EXERCISES

For the following questions, identify a single facility environment in which you would like to situate your responses and answer each question accordingly. Answer each question in the context of a fire, safety, or security audit.

1. Should audits be scored? Why or why not?
2. What are the pros and cons of an "all or nothing" approach to scoring audits?
3. What are the pros and cons of a scaled approach to scoring audits?
4. How should questions that are not applicable to a given facility be managed in the scoring process?
5. In what ways can calibrating auditors be of benefit when addressing audit scoring?

19 Auditor Selection and Training

AUDITOR SELECTION

The auditor selection process is an opportunity to identify the individuals who are most qualified and possess the skills that are necessary to be part of the audit team. The first step in auditor selection is to engage with Human Resources to determine the internal selection and hiring processes that are in effect within the organization. Labor law and company policies will dictate how positions must be posted and filled within an organization. The Human Resources Department will be the authoritative group to provide guidance on how to post auditor positions, interview, and select candidates. They will help create a legal job description, post the position, and manage the selection process.

JOB DESCRIPTION

In order to post an auditor position, it is necessary to write a concise job description. This should be a description of what an auditor will do, including work activities and the amount of travel. The job description should provide an interested person with a clear understanding of what the job will entail.

In addition to an explanation of the position, two issues must be considered when creating a job description. First, position requirements must be established. Items that are included on this list must truly be considered requirements and can be defended as such. Human Resources personnel can provide guidance on how to properly write this part of the job description. It will be a team effort in that Human Resources can aid from a labor law perspective, while auditing professionals can provide practical information on performing the tasks of an auditor. Some requirements may be preferential, such as possessing a bachelor's degree in a related area of expertise (fire, safety, or security), while others may be driven by the nature of auditing tasks, such as being able to climb ladders while conducting a physical inspection of a facility. It will be important to partner with Human Resources personnel on position requirements to ensure compliance with laws, such as the Americans with Disabilities Act, and to ensure the list truly represents requirements, as team members will be selected based on their satisfaction with each item on the list.

The second issue to consider in writing a job description is listing candidate's preferences. For example, it might be preferred that candidates speak Spanish due to the demographics that exist in certain facilities within the organization. Though it may be preferred that candidates speak Spanish, this is not a requirement that would eliminate someone from the auditor selection process.

 DOI: 10.1201/9781003371465-24

JOB POSTING

Once a job description has been created, it must be posted for candidates to read. This can be done using numerous avenues, such as:

- Social media platforms
- Organizational website
- Employment websites
- Professional publications

Two issues that can dictate where a job is posted are cost and exposure. All positions posted within an organization will typically be managed through Human Resources. A budget may be in place to provide funding for staff recruitment. The cost of recruiting that is incurred through job posting may be closely managed to ensure the money is being spent wisely based on the specific jobs being posted. The organizational website operated by the company is one area that involves little or no expense. However, an advertisement in a professional publication could be costly.

Exposure to job postings is determined by evaluating the degree to which the target audience of the posting is being served. An evaluation of the market for social media platforms might prove to be too general, whereas an employment website may prove to be ideal due to the volume of individuals navigating such tools in search of auditing employment opportunities. An organizational website may also prove to have the proper exposure due to the potential of identifying internal candidates who routinely search for advancement opportunities.

AUDITING SKILLS

Consideration should be given to the skills that auditors must possess. Conducting an audit requires the use of numerous skills that must be adapted to fit a given environment and moment in time. A timeline can be used to establish skills that are needed throughout the audit process.

- *Organizational Skills*: Scheduling for and planning an audit requires organizational skills. The auditor will need to plan travel, lodging, and other logistical considerations for the audit. Communications will need to be sent in advance of the audit to prepare the facility and state what can be expected throughout the process. The auditor will need to maintain an organized flow throughout the audit to eliminate confusion. Information from the audit will need to be organized into a clearly written final report that establishes the follow-up activities that will occur.
- *Time Management*: The auditor will need to manage time from the moment the audit begins. The three core phases of documentation review, facility inspection, and employee interviews must be taken into consideration when conducting the audit on-site. Failure to allow sufficient time for each activity can result in the audit not being fully accomplished. Time can also become an issue if events arise that require an alteration of the timetable,

such as a Maintenance Manager being on vacation for a day who was previously included in the audit process. Adapting the schedule to accommodate such random occurrences will need to be effectively managed. Following the audit, the auditor will need to continue to manage time by submitting a final report within a specified timeframe. Allowing too much time to pass before submitting the final audit report may reduce the degree of focus on the audit event by those who received the audit.

- *Analytical Skills*: The auditor must be able to analyze what is seen at the facility, considering the requirements contained in the audit document. Processes and procedures must be analyzed and properly interpreted based on what is asked for in various sections of the audit. This will require the auditor to become increasingly familiar with the organization and the work processes that occur.
- *Technical Skills*: Technical proficiency must be achieved by the auditor in the area in which the audit is being conducted. For example, if a security audit is being conducted, the auditor must be technically proficient in the field of security to include such things as loss prevention, legal liability, vulnerability assessment, and contract management. Technical proficiency will help to build the credibility of the audit process by having subject matter experts conduct audits in their respective fields.
- *Verbal Communication Skills*: During the audit, the auditor will be in contact with a large spectrum of people at a facility. This will include production workers, maintenance staff, supervisors, managers, and the facility manager. Communication skills are critical to being able to effectively interact with individuals in various levels of the organization. Communication skills will be instrumental in building trust with facility personnel.
- *Written Communication Skills*: The auditor will need to develop a formal written report following the audit that communicates the level of performance of the facility. This will require the auditor to utilize technical writing skills to communicate the issues that were identified, the relevance of the findings, and the level of priority that each deficiency must be assigned. The auditor will also need to communicate effectively via e-mail prior to and following the audit in a manner that maintains a constructive relationship with facility personnel.

This list of skills provides an overview of things to consider when selecting auditors. Consideration can be given to those skills that a candidate already possesses and those that may be developed.

AUDITOR TRAINING

Once auditors have been selected, they must be trained to effectively carry out the audit process. A challenge in this area is that the word "training" can mean different things. It can mean:

- Having auditors experience a Power Point presentation

- Conducting a practice audit
- Navigating a computer-based training system
- Engaging a consultant to direct the training sessions
- Participating in role-playing exercises

The challenge is to determine what training needs to be accomplished within the audit program. Principles of adult education and learning must be understood to properly establish a training program. Rather than simply putting together a training program based solely on cost and content, understanding the principles of adult education and learning will help establish an auditor training program that will be most effective.

THE WORKPLACE

Auditors can learn in the classroom, but they can also learn in the workplace. Jobs held by auditors can have a unique influence on how they develop their perception of fire, safety, and security in the workplace. Work experience in different organizational roles and places of employment can play a significant role in the development of an auditor.

Understanding workplace learning can be beneficial in structuring auditor training due to two issues. First, it can aid in understanding the current condition of auditors based on their previous experience in the workplace. If workplace learning has fostered the internalization of fire, safety, and security values, then training the auditor on these issues may not present a significant challenge. Second, workplace experiences can be used as an opportunity to train auditors. Once auditors are actively engaged in the audit process, training can continue by using the experiences they face as an ongoing methodology for conducting training. Rather than learning stopping after initial training has been conducted, training can be perceived as an ongoing opportunity through the experiences that are encountered in the audit process.

IN THE WORLD

Beyond academic and work environments lie life experiences that may be unique to an individual based on a number of variables, such as social culture, geographic region, and educational background. For example, the geographic region in which an auditor has lived may have a cultural influence on how fire, safety, and security in the workplace are viewed. Those who live or were raised in the West Virginia coal fields may have a strong sense of workplace safety due to the experiences shared by friends or family members, whereas an auditor who lives or was raised in a white-collar environment may have less focus on workplace safety issues due to perceived lower hazards in the workplace. The school the auditor attended may also impact how perceptions of the workplace are developed. The curriculum or region of the country where the school is located may include more or less information related to fire, safety, and security in the workplace.

Adult education research begins to set the stage for viewing the auditor as an adult learner with many variables that can influence learning. Rather than seeing these individuals as employees of a company, they can be seen as unique people who have certain needs and contextual issues that can impact how they learn. They are employees of a company and auditors, but they are also individuals whose learning needs must be understood.

ANDRAGOGY

An evolution in the way to view the adult learner is through the paradigm of andragogy. Andragogy offers principles of learning for adults, similar to the way pedagogy provides principles of learning for children. It is a framework popularized by Malcolm Knowles and, simply stated, is the art and science of how adults learn. The "art and science" aspect of andragogy creates a difficult use of words when attempting to more fully define what andragogy means and subsequently how to apply it. The "art" aspect takes into consideration unique individual and environmental issues that can impact adult learning, while the "science" aspect takes into consideration things that have proven to be beneficial in fostering adult learning.

EXPERIENTIAL LEARNING

Experiential learning is a tool that has proven to foster adult learning. Practical exercises can be designed to expose the adult to the events that will be encountered. In the realm of auditing, trainees can accompany a seasoned auditor and observe how an audit is to be properly conducted. This practical experience allows the adult learner to immediately understand and apply important information in the training process.

STORYTELLING

Stories can have a great impact on the learning process. Fire, safety, and security in the workplace can be explored more deeply through interviews with auditors to determine the degree to which story-telling has been used to shape their perspective of the workplace. Seasoned auditors can share stories with trainees that depict the challenges and successes they experienced when performing audits. Trainees can visualize what occurred as the stories are told and begin to conceptualize how they might respond properly in similar circumstances.

TRAINING ENVIRONMENT AND CONTENT

Information regarding adult learning helps inform our understanding of how training should occur. There is a need to focus on the practical application of information through exercises and support the activity with reasoning as to its importance. It may be necessary to impart basic information through a PowerPoint presentation, but this should not be the sole medium of instruction. Auditor candidates will need to engage in activities that have meaning to them in relation to the audit process.

Experiential learning can be used proactively by establishing scenarios to which auditors can be exposed. Auditor training can utilize the concept of experiential learning in three ways. First, role-playing exercises can be created where the train-ees actively engage in scenarios they may face in the process of conducting an audit. Second, problem-posing scenarios can be used to engage trainees in actual issues that might be experienced so that solutions can be identified. Third, practical skills activities can be designed to give the trainees the opportunity to actually do what is being discussed in the training session. They can be taken on a pilot audit, where they actually conduct all or portions of an audit in an effort to become familiar with how the process functions.

Learning is not limited to the classroom. Informal and incidental learning can be utilized as an ongoing strategic opportunity to train auditors. One way in which this can occur is to take advantage of situations as they arise to address auditing issues, such as a discussion with a plant employee where the trainee is present. These infor-mal and incidental events provide rich opportunities for the trainee to learn from what is being said as well as to make connections between what has been discussed in a classroom and what occurs on the plant floor. Another way in which this can occur is to ask each person in an auditor training class to report back on an event that can be classified as informal or incidental learning that they have experienced throughout the week, such as a discussion with a family member or something they viewed on television. Learning opportunities abound in our daily environments, and these events can be incorporated into an auditor training program.

Self-directed adult learners need to have control over their learning. This prin-ciple of adult learning can be challenging for a trainer because it challenges the paradigm that is used to construct many training programs. Typically, a training program may be designed by conducting a needs assessment to determine the gap in knowledge, employing a subject matter expert to create training content, and then having a professional trainer mold the information into a formal presentation. Though this process is typical, it may not be necessarily considered effective due to the lack of control the adult learner has over what is being presented in the training session. He or she simply arrives for the class and is walked through the material. Instead, the adult learner should be given as much control over the process as pos-sible. Rather than scripting the entire training program, the trainee can be provided with choices. For example, an objective of the training program will be to ensure trainees are able to conduct employee interviews. One way this can be done is to create role-playing scenarios in which auditor candidates can act as if they are con-ducting employee interviews. However, some adults do not feel comfortable in the artificial environment role-playing creates and prefer to actively engage in the pro-cess while being observed by the trainer on the floor. Conversely, some adults may feel intimidated by an actual interview with an employee and may prefer the safety of a role-playing environment. Allowing trainees the opportunity to select from vari-ous learning opportunities will provide them with some degree of control over the learning experience.

Banking refers to the process a trainer uses to simply "bank" knowledge in the mind of the trainee by the act of imparting information. The student plays a passive role by absorbing the information with little or no interaction. This is an example of how not to conduct auditor training. Banking is the antithesis of how adult learning should occur, yet it is often utilized due to the convenience that it affords the trainer.

It is easy for a trainer to create training material, bring trainees into a room, dispense the information, and collect documentation of the training through participants signing a training log or taking a brief exam. However, this effort is ineffective in the realm of adult education. Rather than being told the information, adult learners need to be engaged in the process.

Trainees have a life outside of work, and issues there can impact their ability to learn in a training program. Financial pressure that may be mounting due to things such as a divorce, a child going to college, or the loss of a family income can have a significant impact on the performance of an adult in an auditor training session. The trainer will need to be sensitive to events that could negatively impact the performance of an individual in the training program.

Trainees can gain great amounts of knowledge from experienced co-workers. This might be most prominently displayed in the form of on-the-job training. Rather than using on-the-job training as a sole training mechanism, it should be more positively viewed within the context of a practical skills component of the auditor training program. Numerous practical skill opportunities should be made available to auditor candidates due to their need as adults to actively engage in activity. Trainees can work with a seasoned auditor as well as engage with knowledgeable location personnel. Working with a seasoned auditor will provide opportunities to engage in conversation when the auditor candidate can ask questions and learn from someone who has performed numerous audits. Similarly, the auditor candidate can engage with knowledgeable personnel at the facility being audited to gain a greater understanding of the processes that occur and then interpret them in relation to the audit document.

Trainees should be provided with a program that requires effort on their part to navigate. The training program could be designed as something that the adult learner must pass in order to be designated as an auditor. Though auditors have been selected, work can be done with Human Resources to determine an avenue in which fully becoming an auditor is contingent on passing the training program. If the auditor candidate is not demonstrating significant effort in the process, then consideration should be given to the individual remaining on the team. If the auditor candidate is not putting effort into the process, then the result is that the appropriate level of knowledge and skill is not being developed, thus making the individual a potential hindrance in the audit program.

Feedback should be provided throughout the training process in an effort to foster behaviors that are needed among auditors. Throughout the training process, the trainer can provide feedback to the auditor candidate on what is being performed correctly and what has room for improvement. Rather than only correcting trainees when they do something wrong, trainers should also commend them when proper behavior is exhibited in an effort to see them replicate that behavior in the future.

The challenge with auditor training is to incorporate principles of adult education into the training process rather than relying on previous methodologies where information is simply dispensed. Adults have life experiences and motivations that create the need for them to be engaged in the process. This can be achieved with a proper understanding of the principles of adult education and effective planning.

CASE STUDIES

FIRE

Sally has worked for the company for 20 years and is passionate about fire safety. This passion drove her to be part of the industrial fire brigade. As a child, she experienced a fire in her home that resulted in the death of her father. She actively reports fire hazards in the work environment and has earned the nickname "Fire Chief" by her co-workers. She graduated from high school but was unable to attend college due to financial issues. Since working for the company, she has completed five online courses in business administration.

SAFETY

Shelby recently completed her bachelor's degree in safety management and has been a Safety Supervisor at a facility within her company for one year. Her performance reviews are glowing, with numerous positive comments by her supervisor and her co-workers. She tends to accomplish tasks on all projects to which she is assigned, but her only critique has been that she does so at the last minute, which has caused some degree of stress for team members.

SECURITY

Allen is a retired police officer with extensive experience in white-collar crime and community policing. His career began as a patrol officer, and he completed his police work as a Captain. He led an effort in a business district within the community that reduced crime by 45%. He has a high school diploma and holds numerous law enforcement certifications.

- If these individuals became applicants in your audit program to become auditors, what are the auditor selection issues in each of these situations?
- Responding to each scenario, would you select the individual to be on your audit team? Why or why not?

EXERCISES

For the following questions, identify a single facility environment in which you would like to situate your responses and answer each question accordingly. Answer each question in the context of a fire, safety, or security audit. Answer each question based on a comprehensive audit rather than an operational audit.

1. What things would you list as requirements for an auditor position posting? Why?
2. What things would you list as preferences for an auditor position posting? Why?

3. What are the top three skills you would look for in an auditor candidate? How would these skills apply within the context of your audit program?
4. What three principles of adult learning will be most useful to you in creating an auditor training program? Why?
5. Based on the principles of adult learning, how would you structure a comprehensive audit training program?

20 Audit Logistics

Writing an audit program requires a great deal of thought and consideration to plan all of the details that will be required to implement the program. Similarly, a great deal of thought and consideration will need to be given to planning the logistics of an actual audit. Logistical considerations require detailed planning to ensure that the audit is carried out as smoothly as possible. In general, the following logistical issues will need to be considered:

- Travel
- On-Site Planning
- Audit Team Location

TRAVEL

Travel arrangements will need to be made for the audit. Though sometimes delegated to administrative staff, the auditor may also need to be prepared to make travel arrangements. Travel arrangements will typically need to include three categories of logistical planning:

- *Airline Tickets*: If the audit is at a distance from the home office, airline tickets will need to be purchased. Booking airline tickets will require consideration as to the defined start and stop times of the audit. If the audit will begin at 8:00 am on Tuesday and conclude by 4:00 pm on Thursday, it is advisable to schedule air travel for Monday that allows enough time to arrive at the destination with a designated amount of time to accommodate potential delays in flights or missing a connection. The weather conditions in certain areas of the country during different times of the year should also be taken into consideration, which could result in possible delays. For example, if a certain airline ticket includes a connection in Minneapolis in January, airplane deicing procedures may result in a delayed departure. Allow enough time to accommodate potential issues that could delay a set flight schedule.
- *Rental Car*: A rental car might be needed to travel from the airport to the facility being audited. Typically, rental cars can be easily retrieved at the airport, which allows for an easy transition in mode of travel from the airport to the highway.
- *Hotel*: Hotel reservations will need to be made as audits typically last more than one day. Hotel selection should be made based on several variables, including access to the Internet, cost, quality of accommodations, and proximity to the facility being audited. In keeping with personal wellness, the presence and quality of fitness facilities in the hotel selection might also be considered.

DOI: 10.1201/9781003371465-25

If working for an organization as an internal auditor, it will be important to understand the policies that are in place that govern travel. Issues covered in a travel policy might include the following:

- *Food*: Some organizations might have a per diem policy for eating meals each day that allots a certain amount for each meal. For example, $12 might be allotted for breakfast, $15 might be allotted for lunch, and $25 might be allotted for dinner. The total dollar amount of $52 per day might be considered a standard rate and could be increased if the travel destination is in a particularly expensive area of the country, such as a major metropolitan area. The per diem amount will be included in travel reimbursement as a flat amount. A second option is open-ended spending. In this environment, the auditor is given personal responsibility to spend what is reasonable and customary based on the region of the country. Exceptions might be made in a company policy for things such as group meals. Generally, the senior person at the meal is expected to pay the bill, which would be taken into consideration when expense reports are filed. When group meals occur, the names of those present will typically need to be captured as a record of the event and to substantiate the amount on the restaurant receipt. The company policy will typically dictate the requirements for what receipts are needed for reimbursement and how they are to be submitted.
- *Selection*: The auditor will have several opportunities to make selections when planning travel for an audit. This will include which airline to fly with, which rental car company to use, and which hotel to stay at. Company policy may allow reimbursement for what is normal and customary for a given trip. For example, a round-trip flight from Minneapolis to New Orleans may consistently cost $1,000 for travel during the business week, while a round-trip flight from Minneapolis to Chicago may cost $400. The cost of each of these flights is dramatically different, but the company may realize that each cost is reasonable and customary. The auditor will need to exercise ethical decision-making in these choices. Certain choices can be made that benefit the auditor. For example, the auditor might be an elite member of a certain hotel chain rewards program and may naturally book travel reservations with that chain in order to personally gain account points for dollars spent on the stay that can later be redeemed for personal rewards. Such a reservation might be made even though a reservation at a comparable hotel is cheaper per night. The auditor will need to be careful to both comply with company policy and exercise ethics when making these decisions.

ON-SITE PLANNING

Arrangements need to be made with personnel at the facility to accommodate various aspects of the audit. This will include things such as:

- *Auditor Workspace*: An area will need to be provided that allows the auditor or audit team to work as comfortably as possible. Due to office space being limited in many facilities, this work area could be something as simple as a table in a conference room. The work area should allow a degree of confidentiality so that conversations can be had regarding the progress of the audit. Access to the Internet should be provided to facilitate the acquisition of information that might be needed throughout the audit as well as the potential for communication via e-mail.

- *Opening Conference*: A meeting area should be defined for an opening conference that will accommodate a small group. Those in attendance should include anyone who has an interest in the audit process, such as the Plant Manager, Safety Manager, Security Manager, Maintenance Manager, Shift Managers, and Department Managers. An area such as a conference room might be necessary due to the size of this group.

- *Employee Interviews*: The logistical considerations of employee interviews will be based on whether employees will feel more comfortable speaking with an auditor on the production floor or in a more private area that is away from co-workers. If interviews are to be conducted on the production floor, they should be done in an area that is not in the primary flow of production. This will ensure the safety of the employee and the auditor. Though the interview is being conducted on the production floor, it is important to consider safety for the event so that the employee can concentrate on the questions being asked and not be impacted by activity in the work environment. Noise is also a consideration. The production environment can contain work processes that generate a great deal of noise, so the interviews will need to be conducted in an area that fosters communication between the employee and the auditor. If the interviews are to be conducted away from the production floor, a room will need to be identified where the employee and auditor can freely communicate.

- *Closing Conference*: Similar to the opening conference, plans will need to be made as to where the closing conference will occur. It is possible that the population for this conference will be much smaller, as those present might be only those who have a primary interest in the audit results. This might include the Plant Manager, Safety Manager, and Security Manager. An area such as the Plant Manager's office might be sufficient due to the smaller size of this group.

AUDIT TEAM LOCATION

The auditors will need office space that facilitates their work activities. Whether they are full-time auditors based in a corporate office or are auditors that have been selected from existing fire, safety, or security personnel, their office space and equipment need to provide an environment that accommodates the work to be accomplished. Below are things that should be considered for the location of the audit team:

- *Technology*: Auditors will need to have access to technology. This might include updated laptops that will be needed for field work, dependable access to the Internet, presentation software and equipment, and cell phones. The ongoing evolution of technology requires an audit team to stay abreast of what tools are available and select those that make sense to accomplish the goals of the audit program.

- *Confidentiality*: The location of the audit team should accommodate confidentiality. The audit team will generate information that is sensitive in nature. They will have e-mail, telephone, and face-to-face conversations with individuals regarding audits that will require confidentiality. Reports produced will need to be maintained confidentially. Each of these issues will require auditors to have a workspace that is free from the intrusion of those who are not directly engaged in the program.
- *Workspace*: The workspace will need to be conducive to producing audit reports and following up on past audits. From a mechanical perspective, this might include ergonomic issues such as the area being well lit, office furniture being properly adjusted to accommodate the physical stature of the auditors, and noise levels being controlled. From a production perspective, this might include properly sized desk surfaces, office supplies, and file cabinets to archive hard copy material.

Whether it is planning travel for an audit or arranging auditor workspace, logistics is a critical issue to be considered within the scope of an audit program. Logistics requires a great deal of thought and attention to detail to execute an audit program in the most efficient manner possible.

CASE STUDIES

FIRE

Al has accepted the position of fire department auditor. He is responsible for conducting annual audits at each of the department's five stations across the city. Upon his arrival at the Main Street station, he is escorted to his work area. He is placed at a desk that is in the corner of the Chief's office. Al is a little uncomfortable with where he will be working over the next three days.

SAFETY

Melissa has arrived at the plant to conduct the annual safety audit. Management has confirmed with her that employee interviews will be performed on the production floor. She decides to integrate the interviews with her facility inspection. While walking through a production area, she randomly selects an employee to interview. She approaches the employee and asks if it is okay for her to interview him. He replies that he would be glad to participate in the interview. He continues by saying that the Department Supervisor made all the employees aware that the auditor might stop in the department to interview employees but that they could not leave their work area and must continue operations. Melissa conducts the interview in the best manner possible given the environment.

SECURITY

Sally is in the process of making travel arrangements for her next security audit. She has identified two hotels that are close to the facility. She is a points-club member for one of the hotels, but it is $10 more expensive per night than the other hotel. Both hotels appear to have similar accommodations. If she stays at the more expensive hotel, she will earn enough points to cover free night stays that will cover her upcoming vacation next month.

- What are the logistical challenges that are present in the scenarios? Why are they a problem? How might the problems be resolved?
- What ethical considerations are important to evaluate?

EXERCISES

For the following questions, identify a single facility environment in which you would like to situate your responses and answer each question accordingly. Answer each question in the context of a fire, safety, or security audit.

1. What issues should be considered when booking airline tickets for an upcoming audit?
2. What company policies might impact travel for an audit?
3. What role do ethics play in planning travel for an audit?
4. What logistical issues are involved when establishing where employee interviews will take place?
5. Where should an audit team reside in relation to permanent office assignments? What logistical issues should be considered in arranging and equipping office space?

21 Audit Frequency

The frequency with which audits should be conducted at a given facility needs to be based on specified criteria. This applies to both basic and complex audits. Where basic audits might be driven by ongoing operational requirements, complex audits will require the evaluation of a number of variables to determine whether the audit should be conducted once each year, once every two years, once every three years, or on a more infrequent basis. In general, the following issues will need to be considered when determining the frequency of complex audits:

- Complexity of the operation
- Organizational risk of loss of the facility
- Facility performance in the area being audited

COMPLEXITY OF THE OPERATION

The complexity of the facility will be necessary to consider due to the potential exposure of the sheer volume of product, equipment, building construction, and spectrum of operational activity. The size of the facility might naturally present issues, such as safety. A large facility could present complex issues that include:

- Handling chemicals
- Working from heights
- Operating forklifts
- Entering permit-required confined spaces
- Welding
- Working on powered equipment
- Managing the storage of combustible products

The volume of this type of activity that occurs at a large facility could dictate that an audit occurs on an annual basis. However, the risk might not be as great at a smaller facility where less of this activity and exposure occurs, requiring it to be audited once every two or three years.

A facility might also be considered complex due to the volume of product that is received, shipped, or manufactured. The presence of a large volume of desired product could make a facility a target for theft. For example, a facility that distributes electronic merchandise might present significant security concerns due to the potential for both internal and external theft. This may cause the facility to be audited on an annual basis, compared to a facility that produces or processes less product being audited once every two or three years.

Another way of viewing complexity is through the number of employees. A facility that employs 2,000 people might contain greater risk or exposure to loss

DOI: 10.1201/97810033714653-26

compared to a smaller facility that employs 100 people. This is not to say that the smaller facility is less important, but that it may simply entail less risk due to the type and volume of exposure.

RISK OF LOSS

Business Continuity Planning is an evolution in loss control that is closely aligned with organizational Emergency Response Plans. Business Continuity is the science of evaluating the supply chain or organizational operations to determine how business can be maintained, or continuity achieved, if a given facility within an organization is lost. Business Continuity Planning typically involves the three stages listed below:

- *Emergency Response*: Risks to a facility must be evaluated, and then emergency response procedures must be established to respond to each risk. Risks for a given facility might include loss of the facility due to fire, flooding, severe weather, workplace violence, and terrorism. The risks must first be identified, and then mitigating procedures must be put in place to respond if they actually occur. This might include something as simple as defining employee evacuation collection areas or something as complex as how to ensure the operation of the sprinkler system.
- *Crisis Management*: Once emergency response procedures are underway, crises that arise in conjunction with the event must be managed. This would include how to manage and triage those who are injured, manage the media, and communicate with family members of managers and workers.
- *Business Recovery*: The goal of the process is to identify how to restore normal business operations if the various risks occur. Secondary procedures would need to be drafted on how to maintain production if a given piece of equipment or area of the building was lost. Contractors would need to be identified who could provide immediate support to restore operations, such as heavy equipment vendors/operators, salvage companies, and electricians. Plans would be written that indicated how each of these resources would be utilized in the recovery process to ensure minimal downtime from the disaster.

The process of building a Business Continuity Plan will help identify how integral a given facility is to the operation of the organization. A facility that is identified as being of critical importance might be audited every year, while a facility that is, to varying degrees, less important might be audited every two or three years. The process of evaluating the risk of loss of the facility will help determine the degree to which it is important to the organization. This risk assessment will help categorize facilities in relation to their importance in maintaining the operation of the organization.

FACILITY PERFORMANCE

The performance of a facility in the area being audited could influence the frequency with which it is audited. In the area of workplace safety, a number of metrics could be identified, such as:

- Increase from previous annual audit score
- Injury rate improvement
- Workers' compensation cost reduction
- Behavior-based safety observation data
- Safety improvement suggestions that have been implemented
- Safety training that has been conducted

The facility can be evaluated based on its performance in a number of areas to determine if it has met a predetermined standard of performance that would qualify it for exemplary status. This achievement could result in the facility being audited on a less frequent basis. For example, over a period of two years, a facility has been able to maintain an audit score in the high 90 percent range, which was preceded by a substantial improvement in the audit score over previous years. This data can be accompanied by a consistent demonstration of improvement and maintenance of performance in other areas:

- Consistency could be demonstrated through reduced injury rates that have been maintained below the national average, as indicated by the Bureau of Labor Statistics
- Worker's compensation costs could be shown as consistently being reduced as a result of fewer and less severe injuries occurring in the facility
- Behavior-based safety data may indicate the percentage of safe work behavior has consistently increased over previous years
- Management and the Safety Committee could demonstrate a history of working together to generate and implement safety improvement recommendations
- An evaluation could be conducted of safety training efforts to determine the percentage of required training that has been conducted

A potential concern with establishing a system of decreasing audit frequency based on measures is that authentic improvement has not been achieved and that the data is being manipulated for a favorable outcome. Though this is a valid concern, utilizing multiple sources of data can reduce the likelihood of it being an issue. Evaluating multiple sets of data over a period of time can help provide assurance that the improvement that is indicated is legitimate.

Establishing audit frequency is a critical component of an audit program. Variables related to the audit scope will need to be evaluated to substantiate the reasoning for the frequency with which audits are conducted. Examining primary variables, such as the complexity of the facility, the risk of loss to the organization, and facility performance, will help to establish an understanding of the need to audit various facilities on an annual or more infrequent cycle.

CASE STUDIES

FIRE

Nancy, the city Risk Manager, has been evaluating the cost associated with safety management and the risk involved with each city facility. She has concluded that the safety budget must be cut by 10% to assist the city in meeting its annual budget. She has communicated to leadership personnel in the Fire Department that one way in which this can be achieved is to evaluate the frequency with which each station is being audited. Nancy believes that the time and money spent in this area for the benefit received could be better utilized in other areas of the department.

SAFETY

Tom has been auditing manufacturing facilities within his organization for 10 years. He has seen a significant improvement in the performance of the organization due to the effort that has been placed into correcting audit deficiencies and implementing best practices that have been identified among all the facilities. Auditing comprises approximately 90% of Tom's job. Due to the improvements he has seen, Tom believes that auditing every facility on an annual basis is no longer necessary. Though he runs the risk of his responsibilities changing, he feels compelled to communicate to upper management that safety audits should be evaluated in relation to their frequency.

SECURITY

Seth has been tracking inventory control reports to quantify how much merchandise might be leaving facilities because of theft. While doing this among the organization's eight distribution centers, he realizes an alarming increase in lost merchandise at three of the facilities. Security audits have been conducted at each distribution once every two years, but Seth realizes the need to increase the audit frequency at these three facilities. He also knows the organizational budget is tight, and it will be difficult to convince upper management of the need to spend more money on the audit program.

- What process can be utilized to evaluate the frequency with which audits are conducted?
- Is budget a viable issue to evaluate when considering a change in audit frequency?

EXERCISES

For the following questions, identify a single facility environment in which you would like to situate your responses and answer each question accordingly. Answer each question in the context of a fire, safety, or security audit.

1. Should all facilities in an organization be audited at the same frequency? Why or why not?
2. In what ways might the complexity of a facility impact audit frequency? In what ways can complexity be evaluated?
3. In what way might Business Continuity Planning be utilized in determining audit frequency for a given facility?
4. What performance variables can be considered in determining the frequency with which audits should be conducted at a given facility?
5. What risk might be involved in lessening the frequency of an audit?

22 Creating an Audit Program

A well-written audit program will serve as a road map to ensure focus remains on the intended purposes of the audit and those purposes are achieved. It will serve as the authoritative document for the audit process and will help everyone associated with the process, both the audit team and those being audited, to know what is expected throughout the process. A written audit program should be composed of the following elements:

- Purpose
- Scope
- Implementation
- Audit Frequency
- Responsibilities
- Procedures
- Audit report
- Scoring methodology
- Correcting deficiencies
- Auditor selection and training
- Annual review

PURPOSE

The written audit program should begin with a clearly articulated statement of purpose. Be very specific as to what is intended to be accomplished by the audit program. Though this may appear to be a fundamental step, it is recommended to gather input from a variety of individuals so the statement will be clear and concise, yet inclusive of all of the issues that need to be addressed. For example, the general purpose of a workplace safety audit is to improve the degree of safety for employees. A good way to articulate this might be to state the following:

Good – The purpose of this audit is to identify at-risk situations to provide a safer work environment for employees.

Though this information may be correct, greater detail could be provided to assist those reading the program to know how this audit differs from other initiatives that share this generally stated purpose. For example, many activities occur in the workplace that are designed to "provide a safe work environment for employees." Additional details in the statement of purpose might include:

DOI: 10.1201/9781003371465-27

Better – The purpose of this audit is to identify at-risk situations to provide a safer work environment for employees. This will be accomplished through a significant event at each facility that will include a review of written programs and supporting documentation, a physical inspection of the facility, and conducting employee interviews.

This statement of purpose provides additional detail that begins to separate it from other workplace safety initiatives. However, one challenge with auditing is the tone that is set for the process. Those receiving the audit may view it as an opportunity for others to find out what is being done wrong at the facility. The statement of purpose provides an opportunity to set a positive tone by addressing additional details such as the following:

Best – The purpose of this audit is to identify at-risk situations to provide a safer work environment for employees. This will be accomplished through a significant event at each facility that will include a review of written programs and supporting documentation, a physical inspection of the facility, and conducting employee interviews. In addition to identifying opportunities for improvement, the audit will also identify things that are being performed correctly. Rather than simply generating a list of non-compliance issues, both positive and negative findings will be reflected accurately in a final report.

This version of the purpose statement provides the reader with a comprehensive understanding of why the auditing program exists. It also sets the stage for a balanced perception of the audit as yielding positive benefits rather than simply being an activity that will generate a list of deficiencies.

SCOPE

Defining the scope of the audit provides an opportunity to clearly state what areas of business the audit will impact. The audit program's scope is simply its range of coverage. The scope of an audit can include a number of aspects. For example:

- *Topic*: This audit program will address all issues related to employee health and safety. Workplace safety issues will be examined to include all office, production, and contract work that is conducted within the organization.
- *Facilities*: This audit program will apply to all facilities within the organization. This will include all manufacturing plants, distribution centers, regional offices, and corporate offices.
- *Compliance*: This audit program will examine compliance with all applicable regulations promulgated by the Occupational Safety and Health Administration (OSHA). This audit will also include an examination of non-regulated topics that are needed to support the Occupational Safety and Health Management System.

The scope of an audit program, as delineated in the examples above, is more dynamic than simply indicating the type of audit that will be performed. The reader of the program should become fully aware of the application of the audit in each environment and, if applicable, what will be investigated in the audit process.

IMPLEMENTATION

Information should be provided regarding how the audit program will be implemented. If the audit program is a newly formed process, details should be provided that indicate to facilities affected by the audit how they will begin to experience the process. This might come in the form of the following components:

- *Training*: Management affected by the audit will need to be trained on all aspects of the audit process. They will need to be trained on the audit program, the audit document, the audit scoring methodology, the audit report, and the audit follow-up requirements. They will need to be made aware of who will conduct the audits and at what frequency they will be conducted. Management will also need to be made aware of the timeline that has been established to implement the audit.
- *Trial Audits*: Initial audits may need to be conducted for informational purposes only. These trial audits will provide an opportunity to correct any deficiencies in conducting an audit and will provide those being audited with the practical experience of receiving an audit. This will help to both increase the quality of the audit process and resolve any fears that those being audited may have regarding the process. Trial audits can be conducted by randomly sampling facilities and then communicating the results of the process throughout the organization. Following this exercise, auditing can then begin as an authentic process.
- *Timeline*: Facilities should be provided with sufficient notice of the audit program to allow them to effectively prepare for their first audit. This will help to build credibility and trust in the process by all who are involved. Rather than it being a sudden requirement that must be met, facilities should have enough time to review the audit and understand the metric by which they will be measured. This will allow them to review their internal systems to identify where opportunities exist for improvement and prioritize efforts to correct issues prior to the audit. The result can be a sense of fairness that will help to develop positive perceptions during the launch of the audit program and in a facility's ability to effectively prepare for the first audit.

The implementation portion of the written audit program is an opportunity to begin building a sense of trust and value in the audit program. Providing facilities with detail in this section will help them to clearly understand how they will be transitioned into the requirements of the audit program. Expectations of facilities must be realistic and achievable in relation to the staffing that can be dedicated to preparation efforts and the risks covered within the scope of the audit.

AUDIT FREQUENCY

The audit frequency will need to be determined based on a number of factors that impact the operation of a given facility. Intervals that might be considered include the following:

- *Daily*: targeted audits, such as housekeeping audits, that are designed to ensure an ongoing state of compliance
- *Weekly*: audits that are designed to address departmental-level compliance, such as safety audits conducted by a supervisor or manager
- *Monthly*: formalized system audits that address the implementation of programs at a facility level, such as a security audit to determine if all loss control systems are operating as intended
- *Annually*: a comprehensive audit that addresses a broad spectrum of issues to include management systems, general compliance, training verification, employee interviews, facility walkthrough, and program implementation documentation; this audit may be extended to once every two years or every three years based on a risk analysis of a given facility

Daily, weekly, and monthly audits are considered basic or operational audits. Such audits are designed to ensure that a given program is being implemented and is carried out according to the initial training that has been provided. They are also designed to ensure that employees and managers maintain the level of performance that has been communicated in training sessions regarding the given program. These audits are typically conducted either by fire, safety, and security professionals within the organization or by managers who have been properly trained to conduct the audits. The time required to conduct these audits may range from several minutes, such as in observing a permit-required confined space entry, to several hours, such as in conducting a monthly security audit.

Annual audits are considered complex and will typically be much more detailed, requiring one or more days to complete depending on the scope of the audit and the nature and size of the facility. Conducting a thorough documentation review, facility inspection, and employee interviews can be time-consuming, thus requiring a number of days to complete the comprehensive annual audit.

A risk analysis must be conducted to determine the frequency of various levels of audits. For example, if a permit-required confined space program is being initiated, the risk involved may require auditing that program daily during the initial phase of implementation. Once it appears that all employees and managers who are engaged in the program are completely aware of the requirements and are executing them in a safe manner, the audit frequency might be reduced to a weekly basis. As compliance and safe work behavior continues to be demonstrated, it can be integrated into the standard monthly process. Due to the risk involved, an ongoing audit of the process might occur by conducting periodic permit-required confined space entry observations. The program would also need to be integrated into a comprehensive audit of the program as part of an annual safety management system audit.

A risk analysis will also need to be performed at the facility level to determine the frequency of complex annual audits. There are a number of variables that could impact the frequency at which these audits are conducted, such as the following:

- *Facility Size*: The size of a facility could dictate the level of risk to be considered. For example, a facility employing 1,000 employees could dictate an annual audit as opposed to every two or three years due to the number of employees that are at risk.
- *Facility Location*: A facility located in an urban area may present more risk when compared to a facility in a rural area.
- *Production Activity*: The type of work performed in each facility could dictate a high risk, such as the manufacturing of hazardous materials or the handling of valuable products.
- *Importance in the Organization*: A given facility may represent a great deal of importance in the supply chain or in maintaining business continuity for an organization.
- *Performance on Previous Audits*: A given facility may have consistently scored in an exceptional range on previous annual audits, so the decision could be made to lower the audit frequency to every two years. The assumption is that the facility has reached an authentic level of high performance and the annual audits are no longer necessary to ensure compliance.

The dynamics of an individual organization will dictate the degree of priority placed on various facilities. It will be necessary to establish criteria within the audit program against which each facility will be measured. This will help to establish a system free of bias to identify the frequency at which each facility will be audited. Controversy may surround this selection process, so it will be necessary to identify a prioritized system based on risk that can substantiate the frequency at which a given facility is to be audited.

RESPONSIBILITIES

Responsibilities should be delineated for all individuals associated with the audit program. An evaluation will need to be conducted to determine the role of each individual in the audit program and what responsibilities each must hold. Individuals associated with an audit program and their responsibilities could include the following:

- *Employees*: engage in the audit process as needed to perform activities such as escorting an auditor through a department, responding to questions regarding work processes, allowing an auditor to observe him/her while performing work, and participating in interviews
- *Supervisors*: engage in the audit process as needed to perform activities such as escorting an auditor through a department, responding to questions regarding work processes, allowing an auditor to make work activity observations, and conducting departmental audits

- *Managers*: engage in the audit process as needed to perform activities such as escorting an auditor through a department, responding to questions regarding work processes, allowing an auditor to make work activity observations, conducting departmental audits, and assisting in scheduling employee interviews
- *Maintenance Manager*: engage in the audit process as needed to perform activities such as escorting an auditor through a department, responding to questions regarding work processes, allowing an auditor to make work activity observations, conducting departmental audits, and responding to audit findings by correcting physical and program deficiencies
- *Fire, Safety, and/or Security Professional*: administer the site daily, weekly, and monthly audit process, act as the liaison in the annual audit, coordinate follow-up activity on all audit deficiencies, and train location personnel on the audit process and individual responsibilities
- *Plant Manager*: demonstrate support for the audit process through communications via e-mail notifications and meetings as appropriate
- *Auditor*: conduct audits on the determined cycle, provide the targeted audience with audit results on a timely basis following the audit, and ensure audit findings are brought to closure through a defined cycle of follow-up

The value of including responsibilities in the written audit program is that it establishes a source of accountability. Individuals associated with the program will have a clear source of communication for what is expected of them in the audit program. Accountability can then be managed by addressing performance based on the degree to which each individual met the expectations of the program.

PROCEDURES

The procedures section of a written audit program will be an extensive description of all the activities that must occur within the scope of the program. Details will be provided as to how the audit program is to be carried out. Procedures delineate all activity that will occur within the scope of the audit program. This section will serve as a road map for carrying out the audit process. Core elements that should be addressed include the following:

- *Implementation*: The process used to implement the audit should be described. A timeline should be provided that delineates various milestones in the implementation of the audit program. This will include a list of dates that will provide managers with adequate time to prepare for the initial audit.
- *Audit Document*: The audit document should be described in a manner that provides managers with sufficient information to know how the audit will be conducted and what instrument will be used to conduct the audit. The actual audit document should be referenced or provided as an appendix to the written program to allow managers the opportunity to explore the document in preparation for the audit.

- *Documentation Review*: A description of the records to be reviewed should be provided. This will allow managers the opportunity to prepare by having all recordkeeping in order at the time of the audit.
- *Facility Inspection*: The scope and detail of the facility inspection should be provided so that managers can allocate enough time and personnel support to address pertinent issues in preparation for the audit.
- *Employee Interviews*: The method by which employees will be selected for interviews should be provided so managers can prepare for the auditor to conduct employee interviews. Information provided could include who will select the employees, where the interviews will be conducted, and what type of questions will be asked.

The procedures identified in the audit program will vary based on whether it is an operational audit (daily, weekly, or monthly) or a complex annual audit. An operational audit will be much more concise due to the nature of the audit, while the annual audit will include much more information.

AUDIT REPORT

The audit report should be described in terms of format and content. This information will provide managers with an understanding of what they can expect to receive following the audit, which will include a score, deficiencies that must be corrected, and identified best practices. Information in this section of the written audit program could include the following:

- *Structure*: A description could be provided of how the audit report will be organized. This could include an executive summary of the audit and sections that address areas of excellence and opportunities for improvement.
- *Content*: Content of the audit report should be described to allow the manager to know what information will be provided in the report. For example, a description of the information that will be included in the report can be provided, such as a list of deficiencies with an emphasis placed on the issue that solutions to these deficiencies will not be provided.
- *Follow-Up*: A timeframe for follow-up by the auditor can be provided to instill a degree of accountability for managers to correct issues found during the audit. Information can be provided that describes the system of prioritized resolution to address findings in the audit.

The audit report can be highly customized to the organization and the type of audit being addressed in the audit program. The objective is to ensure that managers clearly understand what the audit report will include and their responsibilities to address and respond to the information identified in the report. A sample audit report can also be attached as an appendix to the written audit program to assist managers more fully in understanding the content of the report.

SCORING METHODOLOGY

The measurement used to produce a final score for the audit should be included in the written program. Managers will then be able to understand the metric that is applied to the audit and how it is constructed. Detail should be provided that indicates each segment that is scored and how the final score is calculated. This might include an even distribution of weight among the different sections of the audit, or some sections may be weighted more strongly than others. Scoring information can be stated as follows:

> *Good* – A score will be provided that indicates the percentage of correct items found throughout the audit.

Though this does communicate to a manager that a score will be assessed at the close of the audit, it is lacking in detail as to how the score is derived. Additional detail can be provided that further defines the scoring methodology.

> *Better* – A score will be provided that indicates the percentage of correct items found throughout the audit. The score will be calculated through an equal weighting of each section of the audit.

This statement provides additional detail in that the manager is now aware that each section of the audit is weighted equally, but it is still not completely clear as to how the score is calculated. A little more detail is needed to fully articulate the scoring methodology.

> *Best* – A score will be provided that indicates the percentage of correct items found throughout the audit. The score will be calculated through an equal weighting of each section of the audit. Each question is worth a maximum of 10 points that will be assessed as follows:

- 10 points: full compliance has been achieved
- 8 points: a level of compliance has been achieved, but there are some minor issues
- 2 points: a level of compliance has been achieved, but there are some major issues
- 0 points: compliance has not been achieved
- N/A: the question is not applicable to the facility

Factoring out the "N/A" questions, the total points possible for a given section will be determined by multiplying the number of questions that do apply to the facility by 10. The total points awarded will be divided by the total points possible and recorded as a percentage. This will represent the score for that section. This methodology will be utilized for the audit as a whole to determine the final score of the audit. The individual sections to be scored will include:

- Documentation Review
- Facility Inspection
- Employee Interviews

This description provides greater detail that will assist managers in fully understanding how the audit will be scored.

CORRECTING DEFICIENCIES

The method used to correct deficiencies found during the audit should be delineated. This process can vary greatly depending on the nature of the audit and who conducts it. An operational audit that is conducted on a daily, weekly, or monthly basis could have a very short-cycled process, whereas an annual audit could have an elongated schedule. For example, the correction of deficiencies in a daily operational audit could occur at the moment the deficiency is identified. An annual audit may include a much more detailed process of hazard classification for each deficiency that would then be placed in a category of being addressed on a 30-day, 90-day, or one-year timeframe due to the large scope of issues that could be found in such an audit and the level of risk of each finding.

The individual conducting the audit might also impact how this section of the written audit program is structured. If the audit is conducted by a third party, the audit report might simply be delivered via e-mail, downloaded from an online platform, or delivered hard copy to the facility, and then the facility must determine how to establish a mechanism to address deficiencies. Liability issues perceived by third-party auditors may impact the degree to which they will supply recommendations as opposed to only identifying the deficiencies that were found during the audit. If the audit is conducted by internal fire, safety, and security personnel, these individuals might work with facility management to address corrective action. They may have a more vested interest in supplying the facility with recommendations as well as assisting in the process of correcting the deficiencies.

AUDITOR SELECTION AND TRAINING

The process by which auditors are selected and trained should be included in the written program. This will serve two purposes. First, it will add validity to the audit process by indicating that auditors must meet established standards prior to conducting an audit. Rather than the audit program being perceived as an activity in which anyone remotely familiar with fire, safety, or security can become involved, it will establish a benchmark of quality that an auditor must achieve. Second, it will communicate auditor requirements for those who may wish to become part of the audit program. Auditor candidates may be found within the organization, eliminating the need to look outside the organization. These individuals may already have knowledge of facility operations, an in-depth knowledge of fire, safety, and security issues, and the respect of their peers.

ANNUAL REVIEW

The audit program should be reviewed on a periodic basis. This might include an annual review of the program. A written audit program should be viewed as a living document that must adapt as the organization grows. Changes such as the work performed by an organization, construction of new facilities, regulatory changes, and employee populations could have an impact on the audit program that must be considered. An established review process will help to ensure that issues are evaluated and appropriate changes are made to the audit program so that it continues to meet the needs of the organization.

CASE STUDIES

FIRE

Sheila has been promoted to the position of Chief in her city's fire department. She has been in the department for several years and has worked her way through the ranks. She has had the opportunity to hold every position in the department. Though she and her team have risked their lives to help others through the years, she is aware that more attention needs to be given to the fundamental safety of each firefighter, both in the fire station and on emergency response scenes.

SAFETY

Bill is a new production supervisor at a rural grain elevator in the Midwest. He recently completed his bachelor's degree in agriculture management and is excited to be in his first professional position. He also has a couple of fears. He has heard stories and seen pictures of grain elevators that suffered a fire that eventually resulted in an explosion. Some of these events resulted in structural damage to the elevator, while others resulted in the loss of life. As Bill is taken on his first tour of the elevator, he notices grain dust that has collected on top of machinery, on ledges, and on the floor.

SECURITY

Allyson was recently hired as the Security Manager for Intec Distribution. Intec is a logistics company that manages the returns process for five of the nation's largest retailers. On her first day at work, Allyson conducts a security survey to determine vulnerabilities at the facility. Her survey includes a review of both the physical security system, such as security officers and camera systems, and material handling processes that could result in loss through process errors and theft.

- How might a written audit program be of benefit in these situations?
- What might be unique issues of concern when developing an audit program in these situations?

EXERCISES

For the following questions, identify a single facility environment in which you would like to situate your responses and answer each question accordingly. Answer each question in the context of a fire, safety, or security audit. Answer each question based on a complex audit rather than a basic operational audit.

1. Why is it important to have a written audit program in the environment that you have selected?
2. Write a statement of purpose that can be included in your audit program.
3. What are three categories that should be considered when evaluating the scope of your audit? Write the scope of information for each of these areas in relation to an audit program that you would like to develop.
4. Delineate your plan for implementing the audit.
5. Indicate the frequency at which you would conduct the audit and state your justification based on the risks that are present in the facility.
6. What positions are present for which you would need to delineate responsibilities? What responsibilities would be assigned to each position?
7. What procedural elements would you need to address in your audit program?
8. Assuming that your audit is performed by internal staff, how would you describe the process that will be utilized to correct deficiencies that are identified in the audit?
9. Why is it important to address auditor selection and training in the written audit program?
10. Assuming that your audit program will be newly developed and implemented, at what intervals will this be reviewed over the course of the first five years?

Part VI

Audit Pathways

23 Workplace Safety Auditing

Principles of quality management taught us a valuable lesson. If things do not get measured, they do not get done. Metrics provide us with a lens through which we can quantify activity and see events and behaviors that occur within our work environments. The data generated from various metrics can help us prioritize and take action in areas where losses are occurring.

In light of the lesson taught via quality principles, if adequate metrics are not in place to measure loss performance, the loss will continue unattended. Appropriate loss control metrics will allow for the ability to clearly see where loss is occurring and assist in determining where resources to prevent loss can be best utilized.

The implementation of quality processes that are now common throughout industry has established the need to measure different aspects of work that is done to more properly assess operational performance and determine methods of improvement. This need is also inherent in the areas of fire, safety, and security. It is imperative that metrics accurately reflect loss history as well as potential and current behaviors to determine how loss can be better predicted.

Conducting various types of safety audits provides an opportunity to examine this issue. Through the audit process, measures can be examined that are used to assess management system performance.

THE CHALLENGE

The current methods most widely utilized to assess safety performance are focused on measuring failures. Though these lagging measurements do provide some adequate information that can be effectively used to understand organizational performance, they can also be misleading. The challenge at hand is to clearly identify a complete body of metrics that adequately address an organization's total health and safety performance. The metrics can be integrated into a safety audit process to present a holistic picture of what is occurring at a given facility.

An example of this challenge is present in something as fundamental and personal as individual fitness. Bill may awake one morning to get ready for work and realize that the figure staring back at him in the mirror is in dire need of getting in shape. He realizes that all those quick morning breakfast donuts are catching up with him. They have helped him make it to work faster and ease into his morning routine, but they have not been quite so kind to his waistline. He also realizes that he chooses to ride the elevator to his fifth-floor office more often than taking the stairs. Recent memories of becoming winded after taking a single flight of stairs have replaced those of his glory days, when he ran a five-minute mile on the track team while barely breaking a sweat.

DOI: 10.1201/9781003371465-29

The challenge is for Bill to choose the right metrics that will help him monitor improvement as he attempts to alter his lifestyle. Along with several of his co-workers, Bill might choose to monitor how many pounds he loses. It cannot be disputed that pounds lost during a fitness regimen are a viable metric. But it is not the only metric or even the most reliable metric that Bill should focus on in his efforts to get in shape.

The focus on pounds is a typical misdirection when attempting to get into better physical condition. It is important to understand that "getting into shape" is not reflective of weighing less but of improving one's complete health. Setting a goal of dropping pounds that are composed of body fat is a worthy objective, but it should not be the only objective. Other metrics that Bill might consider include:

- Calories eaten each day
- When food is consumed
- Blood pressure
- Body fat percentage
- Improvement in endurance
- Volume of "junk" food reduced
- Saturated fat intake
- Nutritious foods eaten

Losing weight may be a step in the right direction, but it may not be enough alone to improve Bill's health. Muscle weighs more than fat, so he must also consider that exercising will increase muscle mass, which may skew weight loss data. By evaluating his total fitness, he must choose the metrics that will best communicate his level of improvement and impact on his fitness.

The same challenge is present in the business environment when loss control is addressed. The need is to identify what is to be accomplished and then define metrics that most adequately quantify performance.

INJURY RATES

One of the most common workplace safety metrics is injury rates. A key driver of this is the federal government. The Occupational Safety and Health Administration collects data that is input into the Bureau of Labor Statistics database of injury rate information on various categories of businesses. Injury rates are communicated as the number of injuries experienced per 100 employees and are calculated using the following formula:

$$\text{Number of Injuries} \times 200{,}000/\text{Actual Hours Worked}$$

Three pieces of data are required to make the calculation:

- *Number of Injuries*: This number represents the number of injuries experienced in the period for which the rate is being calculated. The period can be a month, a calendar year, or a fiscal year. It can be any specified period over which performance needs to be measured.

- *200,000*: This number is a constant in the formula regardless of the length of the period for which the calculation is being made. 200,000 is the number of hours worked by 100 employees over the course of a year (100 employees x 40 hours per week x 50 weeks = 200,000). This constant converts the end number into a representation of the rate per 100 workers.
- *Actual Hours Worked*: The actual hours worked must be factored in so that they are consistent with the period for which the number of injuries was extracted. Note that this figure is "hours worked" and not "hours paid." Hours paid for time such as vacations and holidays will inaccurately inflate the volume of hours used in the calculation resulting in a skewed rate.

For example, the injury rate of a facility that is being audited for the first quarter of the current fiscal year must be calculated and compared to the injury rate of the first quarter of the previous fiscal year to determine if there has been improvement. In order to do this, the following information must be gathered to calculate the rates:

Period	Number of Injuries	X	200,000	/	Actual Hours Worked	=	Rate
Q1: Current FY	42	X	200,000	/	1,943,863	=	4.3
Q1: Previous FY	40	X	200,000	/	1,564,392	=	5.1

This chart points out important information related to injury rate data. The number of injuries increased between the previous and current quarters, yet the injury rate decreased. This is an indication of why an injury rate is a more accurate metric compared to simply reviewing the number of injuries that occurred. The injury rate is a function of the number of hours worked during the measured period. A simple number of injuries does not communicate that detail. If we were to only look at the number of injuries, concern might be shown due to a 5% increase in injuries. What must be taken into consideration is the hours worked, resulting in a decrease in the total injury rate. In this case, the increase in hours worked in Q1 of the current year resulted in a rate decrease from Q1 of the previous year. A rate will provide a true comparison between time periods and can be used to effectively assess workplace safety performance. This can be done to compare companies, business units within a company, or individual plants. Using the rate as a metric in the example above illustrates that what may appear to be a 5% growth in injuries is actually a 16% reduction in true performance.

The injury rate can also be utilized as a benchmarking tool to compare rates to those of the national average for a given industry. The Bureau of Labor Statistics maintains historical data based on various industries. National injury rates for industries are published in their material, which can be readily found on their website (www.bls.gov). The national average industry rate for a given industry sector might be found to be 6.8. This information provides the ability to compare facility progress in a chart such as the following:

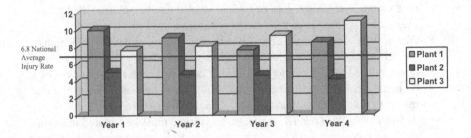

Injury rates can be reviewed in an audit to gain an understanding of incidents that have occurred within the facility over the period being audited. For example, injury rates can be reviewed to determine workplace safety performance as of the time the last audit occurred and compare that performance to the injury rate since the previous audit. The previous year and current year comparisons can provide one perspective of workplace safety performance based on the trending pattern seen in injury rates over a period of time.

HOUSEKEEPING AUDITS

As a result of years of experience in the industry, it is interesting to note employee perceptions when housekeeping is identified as a priority. It is unfortunate that such times may center on site visits by customers or corporate representatives. The Plant Manager wants to make a great impression on the visitors. The order is issued to ensure that all departments focus on housekeeping to make sure the facility is in perfect condition. The visitors arrive, and they make glowing remarks about the condition of the facility. But clutter and disarray begin to work their way back into the system within a week after the visitors have gone. Employees perceive that housekeeping and their resultant safety is only of concern when "outsiders" come to visit. One employee once gave feedback that an evaluation should be done on how much overtime is worked on such issues immediately prior to the annual safety audit.

Housekeeping is an important ongoing process to ensure the safety of employees, contractors, and visitors. From a loss prevention perspective, housekeeping is the root cause of numerous fundamental accidents in the workplace. Poor housekeeping can result in slip/trip/fall injuries from debris left in a walkway to an explosion from combustible dust that is not collected and controlled during production processes.

Housekeeping audits are a form of evaluation that will occur most frequently among the various types of audits mentioned in this chapter. This is due to the ease and speed at which housekeeping problems can materialize. A department may look in great condition one week but can regress to an unsafe environment after a few days have passed. This timeline leads to the need to conduct housekeeping audits as frequently as on a daily or weekly basis.

Keeping objectivity in mind, a document should be created that clearly states the expectations for the department. Each stated expectation should be able to be objectively answered with a "yes" or "no" response. Avoid subjective statements such as "Production areas are clean." The term "clean" can have different meanings for different people. An environment that is considered "clean" by one person may not be considered "clean" by someone else. As opposed to simply stating "clean," the housekeeping audit document should have a number of statements for each house-keeping issue that clearly state what "clean" looks like. This could include:

Housekeeping Item	Yes	No
Aisles are free of clutter.	X	
Boxes are neatly stacked.		X
Floors are swept clean of all debris.	X	
Equipment is stored in its designated area.	X	
Exit doors are unobstructed.		X

This brief example provides ways to specifically communicate housekeeping requirements in a way that limits subjectivity. Placing an "x" in either the "yes" or "no" column provides the user with a scoring mechanism to provide metrics that can be monitored over time. Three "yes" responses and two "no" responses equate to a score of 60% (3/5 = .60). This metric can be evaluated in future weeks to determine if improvement is achieved or if the work area experiences regression. Data showing this performance metric can be converted to a chart such as the following and com-municated to applicable departments:

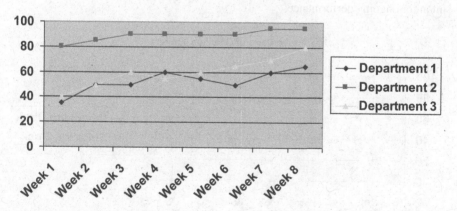

Departmental Housekeeping Scores

ITEM-SPECIFIC AUDITS

It may be useful to identify specific issues that impact workplace safety efforts. Item-specific audits can be developed on an as-needed basis to determine performance in a specific area. An example of this in a warehousing environment could address the need to guard against trailers being pulled prematurely from the dock while work-ers are inside. This type of incident could cause a great deal of loss in the following areas:

- Downtime to repair equipment that may have been damaged, such as extendable conveyors
- Lost production until the damage is repaired.
- Damage to merchandise that has been loaded and falls due to not being secured
- Employee injury

A policy may be in place that requires dock workers to chock trailer tires prior to disconnecting from a tractor. A metric that can be applied to this issue is the percentage of trailers that are found to be chocked during random dock audits. Data can be collected on a form as follows:

	Properly Chocked	Improperly Chocked	Percentage Chocked
Receiving Dock	15	10	60%
Shipping Dock	20	5	80%

Percentages are calculated as:

Number Properly Chocked / (Number Properly Chocked
+ Number Improperly Chocked)

This data from periodic audits can be accumulated and placed into a chart to monitor ongoing performance:

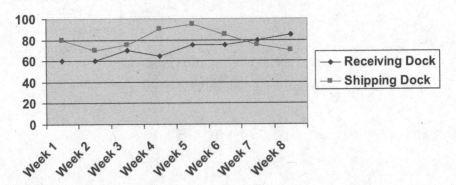

HEALTH AND SAFETY MANAGEMENT SYSTEM AUDITS

The most complex auditing tool available is one that evaluates and quantifies a health and safety management system. This audit breaks down a health and safety management system into primary components and rates each component on an in-depth level. Sections that such an audit will evaluate will include:

- Documentation Review
- Facility Inspection
- Employee Interviews

Additional sections can be added, but these three areas address the macro-level issues that are typically evaluated. Adding other primary sections could cloud the value of an end score to the audit because other items tend to be a subset of these three primary sections. For example, written compliance programs may include the subsets of:

- Written program content
- Implementation activity verified through appropriate documentation
- Employee training

A temptation might be to gravitate toward one sensitive issue, such as employee training, and make it a primary topic. Doing so could skew the final score due to breaking the logical flow of a health and safety management system. All subset items should be identified and placed under one of the three primary headings to maintain an orderly assessment of the system.

AUDIT CONTENT

The first step in establishing an effective system audit is to determine the scope of the content to be evaluated. This issue sets the stage for this type of audit need to be company-specific and, in some cases, site-specific. This also raises the concern of purchasing pre-designed audit systems, whether in electronic format or as services from a consultant. If consultants are used, care should be taken to ensure that a customized audit has been created and implemented. Though consultants are a resource available for accomplishing this type of audit, in many cases, internal personnel are available with the expertise to create the audit document and conduct an audit at different sites within an organization.

Loss control professionals within a health and safety group should be able to evaluate a business and determine the scope of issues that need to be addressed to achieve an effective health and safety management system. Items within this scope will include both regulatory and non-regulatory issues:

- *Regulatory*: One item that might be defined is to implement a Confined Space Entry Program due to exposures to employees who must enter such spaces during production or maintenance activities. This is a regulatory issue and is mandated by the Occupational Safety and Health Administration.
- *Non-Regulatory*: Quantifying and measuring employee safe work behavior may be identified as being needed. This tool is a component of a behavior-based safety initiative and is not regulated by any governmental agency.

A fallacy of many audit tools is that they focus only on the regulatory components of a health and safety management system. Another fallacy is that some may not consider physical condition regulatory requirements as a final scoring metric. This lack of comprehensive inclusion of all issues may lead to inaccurate conclusions based on inaccurate or limited data from the audit.

AUDIT FORMAT

The format of the audit should be arranged to fit the needs of the organization. One tool to accomplish this is spreadsheet software. Using a spreadsheet will allow the flexibility to:

- Format the audit document in a fashion that makes sense for the organization.
- Easily make changes to the document over time.
- Mathematically calculate scores for various audit sections and the audit as a whole.

A series of tabs can be established to represent a cover sheet and the three primary categories of documentation review, physical inspection, and employee interviews.

Web-based and app-based audit tools continue to evolve. These must be explored in great detail to ensure they provide the functionality needed to meet the needs of the employer.

DOCUMENTATION REVIEW

The documentation review section is designed to measure the level of legal compliance achieved by various programs required in the work environment by organizations, such as:

- Occupational Safety and Health Administration
- National Fire Protection Association
- State government legislation
- Local government legislation

Loss control professionals within an organization will be able to evaluate legal requirements affecting the organization and create applicable written compliance and employee training programs. Following is a potential section in the audit document related to one of these topics:

PROGRAM: HAZARD COMMUNICATION

	Yes	No	N/A	Comments
Does the written program include:				
Assignment of responsibilities?				
Method for obtaining safety data sheets?				
Maintaining a chemical register?				
Introduction of new chemicals?				
Primary container labeling?				
Secondary container labeling?				

(Continued)

	Yes	No	N/A	Comments
Has employee training been conducted that includes:				
How to read a safety data sheet?				
How to read container labels?				
Location of safety data sheets				
Potential hazards of applicable chemicals?				
Necessary personal protective equipment?				
Emergency response procedures?				
Has the program been reviewed within the past year?				

Spreadsheet software allows the ability to insert text boxes that provide greater detail regarding individual questions. This gives the opportunity for location personnel to clearly understand what they must have in place to receive full credit for a given question. The last question in the example above states, "Has the program been reviewed within the past year?" A text box may be inserted that more clearly elaborates, "Facility management must provide signed documentation stating who conducted the review, the date the review was conducted, what changes were made, and when those changes were communicated to employees." This will give facility management the understanding that they cannot simply say that they did the review at the time of the audit but that they must supply documentation of the event.

The auditor will place an "x" in the appropriate box for each question. If the question is marked as not applicable, it will not count against the facility in the composite score. Comments can be recorded by the auditor to assist the facility in improving on a certain line item. Positive comments can also be included to identify performance in a certain area as a best practice.

A spreadsheet is capable of counting the "x" volume of each "yes" and "no" response to formulate the total points possible for the facility. This provides for no penalty to be assessed for responses that are not applicable to the facility. The spreadsheet can then calculate the percentage of "yes" responses from the whole group to determine the percentage score for this portion of the audit.

PHYSICAL INSPECTION

The physical inspection section of the audit provides an opportunity to see how workplace safety programs are implemented in the facility. This will require an actual walkthrough to inspect the conditions of the entire facility.

Similar to employee interviews, subjectivity is a potential barrier to conducting a physical inspection and obtaining a usable scoring metric from an inspection of the general conditions of a facility. Subjectivity can be controlled to a great degree by:

- Clearly defining what "good" or "acceptable" looks like
- Calibrating auditors through training and ongoing communications
- Establishing consistent line items to be rated

A section from the physical inspection section of an audit could be as follows:

PLANT AREA – MAINTENANCE SHOP

	Yes	No	N/A	Comments
Are the floors clear of all debris?				
Are tools properly stored?				
Are tools in safe operating condition?				
Are chemicals properly stored?				
Is the drill press bolted to the floor?				
Are safety signs posted as needed?				
Is machine guarding in place as needed?				
Does the eyewash station function properly?				
Is hazardous waste being processed properly?				
Are sprinkler heads unobstructed by storage?				
Are exits clearly marked?				
Are exit pathways clear?				

The auditor will place an "x" in the "yes" box if the question is being executed in an acceptable manner and will place an "x" in the "no" box if the line item is found to be unsatisfactory. "N/A" will be marked with an "x" if that line item does not apply to the facility being audited. A spreadsheet can sum all the boxes marked with an "x" to determine the total possible point for the plant and then calculate a percentage of "yes" responses for the score.

EMPLOYEE INTERVIEWS

Employee interviews are an outgrowth of the written compliance program assessment section. Written compliance programs and associated documentation may indicate that employee training has been conducted and ongoing safety meetings are being held. Though evidence might be represented on paper, employee interviews are critical to determining how well education and training were accomplished. Documentation and communication are of no value unless they are understood, appreciated, and applied by the audience for whom they are intended. In this case, it is employees. An evaluation must be made to determine the scripted questions that can be asked of employees:

- The auditor is provided with a structure by which the interview can be conducted.
- The scoring of responses provides some degree of calibration when evaluating responses among facilities being compared within an organization.

In keeping with the Hazard Communication written compliance program review example used earlier, an organization will need to evaluate the program to determine what questions will lead the auditor through sufficient interaction with employees to determine if they have truly learned and are in the process of applying the material. It is recommended to limit the list of questions to three questions that target the scope of activity that must be verified by the auditor. The primary reason for limiting interviews to three questions is time efficiency in conducting audits that will involve numerous employee interviews. An employee interview section of an audit could be developed as follows:

PROGRAM – HAZARD COMMUNICATION

	Yes	No	N/A	Comments
What can you use to access chemical information?				
How do you go about working safely with chemicals?				
How are new chemicals introduced into the workplace?				

Though these questions are scripted, the auditor can feel free to ask additional probing questions as the interview unfolds. It is imperative that the auditor assess each designated question to ensure calibration among all scores generated from the audit document when used across multiple facilities. It is also important to note that the questions provided in the example above are open-ended. Such questions avoid a simple "yes" or "no" response and cause the employee to respond in narrative form.

The auditor would place an "x" in the "yes" column if the employee's response is believed to be satisfactory. An "x" would be placed in the "no" column if the auditor believes the employee's knowledge to be lacking in that specific area. An "x" will be placed in the "N/A" column if the question is not applicable to employees at that facility.

Subjectivity of the auditor is a potential problem when conducting employee interviews. Though this can never be eliminated, it can be controlled through:

- Scripting questions that must be asked by all auditors
- Calibrating auditors through appropriate training on what responses would be acceptable and those that would be unacceptable

EVALUATING THE SCORE

The metrics of individual and composite scores from a comprehensive health and safety management system audit provide valuable insight into how to identify improvement opportunities for a facility. The composite score provides an overview of their program, while the individual scores provide a specific indication of performance in each area. Below is an example of a score report:

Section	Score
Documentation Review	95%
Facility Inspection	75%
Employee Interviews	55%
Composite Score	75%

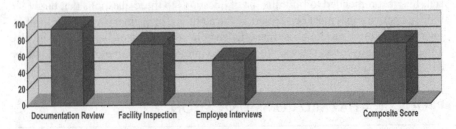

These data indicate that the facility has spent a great deal of time perfecting their written health and safety program material but has been weak on follow-through and implementation. Though written programs are important, it is the lack of employee knowledge and poor general plant conditions that can result in losses for an organization. Written programs are necessary to comply with federal, state, and local regulations. What is written in programs must be blended with accomplishing employee training and improving physical conditions.

Data from these audits can also be graphed to compare facilities within an organization.

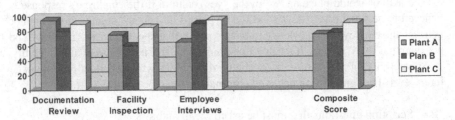

EMPLOYEE PERCEPTION SURVEYS

Employee interviews in a comprehensive health and safety audit will determine what workers know regarding safety issues in the workplace, but they will not necessarily indicate how workers truly feel about their safety. Management can potentially control and manipulate employees to the point where an employee can regurgitate the correct answer to any safety question, but that does not mean the employee has bought into and taken ownership of the safety program.

Employee perception surveys can be developed and implemented as an audit tool to help gain a true understanding of how employees feel about safety in several areas. These surveys may also be referred to as "safety climate" or "safety culture" surveys.

Leadership's knowledge of employee perceptions is critical to understanding how safety efforts are viewed. For example, a manufacturing plant may have had a recent series of injuries and property damage incidents. Management evaluates the situation and concludes that they need to implement a safety incentive program. The incentive program is designed to reward employees based on how many days the plant goes without experiencing an accident. A few weeks later, management is confused when they realize that not only have the accidents continued, but there has also been a slight increase. They are unsure as to how this can happen.

It is important to understand employee perceptions in situations such as this. A primary pitfall with incentive programs is that employees might perceive the effort on the part of management as an attempt to "buy" safety. Employees might naturally respond in a negative manner to what management thought was a positive gesture. An increase in reported injuries can also occur when workers feel more comfortable reporting injuries that would have gone unreported in the past.

Perception surveys are designed to ask an array of questions that address how employees think and feel about different health and safety issues. This could include questions such as the following:

	Strongly Disagree	Disagree	Neutral	Agree	Strongly Agree
My manager cares about my safety.	1	2	3	4	5
I receive sufficient training to do my job safely.	1	2	3	4	5
I am willing to coach a co-worker when they work unsafely.	1	2	3	4	5
I feel comfortable reporting unsafe conditions.	1	2	3	4	5

A comprehensive safety perception survey could be distributed to employees to complete in an environment where they are sure that management cannot see their responses. Confidentiality and anonymity are of the utmost importance. Employees must feel free to respond truthfully to each question without fear of penalty from management.

Surveys can be collected, sealed in an envelope, and sent to an objective third party for evaluation. This third party could be a consultant or an in-house corporate loss control professional. Survey results must be tabulated and then communicated to facility management for review. This will be an opportunity for facility management to see health and safety from the perspective of their employees. Agendas and biases must be left out of the discussion. The information from the survey must be taken at face value. An action plan can then be established that includes:

- Prioritize findings from the safety perception survey.
- Establish a strategy for improvement based on prioritized items.

A safety perception survey should not be a one-time-use product that is focused on for a short term and generates a flurry of activity but is soon forgotten. The audit value of a perception survey is to monitor ongoing performance. The survey can be repeated periodically to determine if improvement efforts have been successful in changing employee perceptions of safety.

GOALS AND STRATEGY

A structured set of safety goals and corresponding strategy will help keep the organization focused on activity. Goals and strategies should be truly significant items that will lead to excellence in safety performance.

GOALS

Auditing goals of a facility will help ensure appropriate items have been identified on which to focus attention and effort. Two issues that all goals should satisfy are:

- *Achievable*: Goals must be perceived by the targeted audience as achievable. The more achievable a goal is perceived, the more effort a group will put into working toward it.
- *Measurable*: A method of tracking progress toward goal achievement should be established to keep the team motivated. Progress should be charted and communicated at frequent intervals to maintain interest in and focus on the goal.

Goals should be high-level items that address significant challenges faced by an organization. Such goals could include:

- Reduce strain/sprain injuries by 20%.
- Conduct safety training among all current employees.
- Reduce property damage costs by 30%.
- Increase safe work behavior in the assembly department by 15%.

These goals are clearly measurable based on specified percentages of reduction or improvement and the volume of training to be completed. The achievable aspect of goals should be carefully determined on a case-by-case basis. Setting a very bold goal to communicate a sense of urgency may discourage the group because they will immediately recognize the goal as unachievable. A more modest goal of reducing strain/sprain injuries by 20% might be more readily received than a 50% reduction goal. A level should be identified that will both indicate significant continuous improvement and encourage the group to work toward it.

STRATEGY

Goals establish where an organization wants to go, but the strategy will include specific action items that will move the organization to its destination. Each health and safety goal should maintain a small number of strategy points that provide practical action to be taken that will influence goal attainment. Potential strategy points are listed below from the list of goals previously mentioned:

- Reduce strain/sprain injuries by 20%.
  - Implement a pre-work stretching program.
  - Train employees on ergonomic risk factors.
  - Train employees on the signs and symptoms of musculoskeletal disorders.
- Conduct safety training among all current employees.
  - Dedicate 1 hour each week to departmental training.
  - Utilize technology and a format that is appealing to employees.

- Reduce property damage costs by 30%.
  - Review incident investigations to determine cause trends.
  - Enlist the feedback of employees associated with the loss to determine preventive measures.
- Increase safe work behavior in the assembly department by 15%.
  - Enlist the feedback of line employees.
  - Provide positive feedback when observing safe work behavior.
  - Identify and remove barriers to safe work performance.

ACCOUNTABILITY

Accountability should be set among everyone at varying levels throughout the organization for safety goals to be achieved. Accountability should be based on strategy and activity, not on the goals themselves. It is unrealistic to hold an individual accountable for a 20% reduction in injury rates across the facility or within a department. There is a broad variety of variables that must be managed to meet a 20% reduction in the injury rate, thus making it impossible to achieve by one person.

Accountability should be set based on what strategic activities individuals can engage in that will influence accomplishing the 20% reduction. Activities to which individuals might be held accountable could include:

- Attend all appropriate safety training sessions.
- Participate in a departmental safety inspection.
- Provide a specified number of safety recommendations for improvement.
- Conduct workplace observations and provide peer feedback.
- Participate on a safety committee.

These activities are proactive and can influence the achievement of a goal. Each of these items is also within the control of an individual to accomplish. A production worker can offer safety suggestions. However, they alone cannot achieve a 20% reduction in the department's injury rate. Providing them with items of accountability that they can control and accomplish can encourage them to become engaged and have a positive impact on goal achievement.

INCIDENT INVESTIGATIONS

Data generated from incident investigations can be formulated into valuable metrics that can be used to assess loss performance in the audit process. Though this analysis is reactive, it is still very important for an organization to learn from and grow from its mistakes. Those who fail to learn from history are destined to repeat it.

INJURY INVESTIGATION METRICS

Information provided from injury investigations can be analyzed and communicated in relation to several issues. Injury investigation data can be broken down into various metrics, such as:

- Type of injury experienced (strain/sprain; burn; impact; puncture)
- Body part injured (head; arm; hand; torso; leg; foot)
- Cause (unsafe environment; lack of training; poor supervisory leadership)
- Cost (worker's compensation; indirect; corrective action)

Information from a selected metric can be gathered and used to generate a graph such as the following:

Injuries Per Department

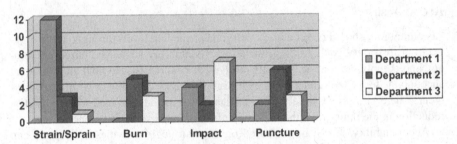

This chart provides a clear comparison of injuries experienced by each department within an organization. High occurrences of injuries can be targeted for safety improvement efforts. A defect in looking solely at this chart for information is that it does not provide an injury rate but simply the number of injuries experienced. However, it does provide trending information that can be evaluated in the audit process.

Property Damage

Incidents involving property damage should be evaluated in two primary areas:

- Causes
- Cost

These two categories are sufficient to help drive improvement through understanding and preventing future occurrences. Causes will help identify what corrective action must be put in place to ensure the event does not recur. It may be found that a given production line routinely breaks down, damaging merchandise and exposing employees to injury. Several investigations may yield the conclusion that there is a lack of a significant preventive maintenance program related to the production line in question. This information may lead to the maintenance department evaluating the work area and instituting preventive maintenance that will stop subsequent potential losses. This information will also cause an evaluation of other similar work areas to ensure appropriate preventive maintenance is in place.

Evaluating cost trends in property damage may lead to where resources should be placed. Cost will need to be evaluated in four areas of occurrence, as indicated by the following diagram:

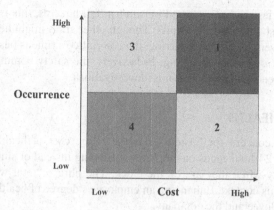

The numbers in each quadrant represent the priority that should be placed on each category of property damage:

- 1: High cost/high occurrence may produce the largest opportunity for loss.
- 2: High cost/low occurrence may not show itself in significant trends among property damage incidents, but in the shear dollar loss per event
- 3: Low cost/high occurrence events may result in small dollar losses in each event, accumulating into large composite losses.
- 4: Low cost/low occurrence may provide little exposure to significant loss.

Viewing property damage loss data in this light will help establish a priority basis on which to address loss.

Near Misses

An event that can be a precursor to an actual loss event is known as a "near miss." Near misses occur frequently in the workplace, and they should be given as much attention as an actual loss event. These are opportunities to receive a warning that a system flaw exists and can be corrected before actual loss occurs. Near misses can be tracked over time to identify trends that can point to primary areas for improvement. Auditing near-miss reports provides an opportunity to evaluate the thoroughness of investigations and the responsiveness of facility management to these incidents.

SAFETY COMMITTEE ACTIVITY

Safety committees are an excellent tool to engage personnel at different levels of an organization in an effort to prevent loss. Just as with any committee, there is the potential for a safety committee to wallow in inactivity, which causes the need for these committees to be audited for their activity and effectiveness. Safety committee performance should be measured by the volume and quality of issues that are identified and brought to closure. The activity of this committee should coincide with data produced via metrics utilized in other areas of loss control. If it has been identified that many injuries are being caused by slip/trip/fall hazards, this is the area where the committee should focus its efforts. Though other areas might be worthy of time invested to prevent loss from occurring, the committee's time is best spent on areas where the bulk of loss is occurring. Proactively, the safety committee could also focus on data generated from near-miss investigations.

EMPLOYEE HEALTH

An increasing concern for businesses is the rising cost of healthcare. This has given way to a 24-hour focus on employee wellbeing instead of simply focusing on employee safety while they are on the clock. The phrase "total worker health" is often used in this context. Enhancing an employee's degree of health will be beneficial to the employee and the company.

- The employee will experience a greater quality of life due to their enhanced health. For example, taking control of obesity may alleviate back or joint problems, and taking control of their diet may assist in controlling diabetes and other health issues.
- The company will benefit from lower healthcare-related losses and increased profitability due to the employee being present and in peak physical condition.

There is a great deal of information available from healthcare insurance carriers that can assist in measuring categories of employee illness. Though medical records themselves are confidential between an employee and the treating physician, the insurance carrier can provide general trending data that is not specific to an individual employee. This data will help to establish what key health issues are affecting the healthcare cost drivers among the employee populace. This, in turn, can help determine potential interventions to stem the occurrence of predominant health issues. For example, a report from the health insurance carrier may indicate that there are many treatments being given for high blood pressure. Facility management may then determine that it is worth the investment to teach proper diet, exercise, and stress management to all employees. Investing in such initiatives can have a return on investment both in the quality of life of employees and in the organization's savings through lower health insurance costs. The audit is a tool that can be utilized to routinely evaluate the effectiveness of workplace wellness initiatives.

CASE STUDIES

FIRE

Barbara is the chief of the city fire department, which includes five stations located throughout the metro area. Though fireground safety is a concern, she also sees the need to address firefighter safety in all areas of the job, including work performed at the stations. The injury rate within the department has been on a slight, consistent increase over the past 3 years.

SAFETY

Fran has been a safety professional for 10 years. She has had a very successful career and is now the Corporate Safety Director for a national construction firm. The organization operates in most of the continental United States and works primarily on large commercial projects but also has a small residential construction group. She has been in the position for 2 years and is now comfortable that the basic safety programs are in place, but she sees the need to implement an annual audit process.

SECURITY

Steve is a Regional Loss Prevention Manager for a growing retail company. He is responsible for 15 stores located in three different states. In addition to his security responsibilities, Steve must also address workplace safety issues at each store. When visiting the stores, he is the subject matter expert on security and safety issues due to the fact that there are no on-site professionals in either of these disciplines.

- What legal issues apply in these situations that could impact the content of the audit?
- What do you believe are the three most effective audit tools presented in this chapter that can be applied in these environments? Why?

EXERCISES

For the following questions, identify a single facility environment in which you would like to situate your responses and answer each question accordingly. Answer each question in the context of a fire, safety, or security audit.

1. How are injury rates calculated? What does an injury rate communicate? How can an injury rate be beneficial in the audit process?
2. What purpose do housekeeping audits serve?
3. Select a single workplace safety issue and identify how an item-specific audit can be beneficial in improving performance in that area.
4. What activity will occur in the three main sections of a management system audit?
5. What are sources of regulations that you must explore when constructing the regulatory compliance portion of a management system audit?
6. What non-regulated topics might need to be included within the scope of a management system audit?
7. While on-site conducting an audit, how would you go about accomplishing employee interviews?

8. What is a safety perception survey? What value does it have in the audit process?
9. When auditing goals, what are two criteria that each goal should meet? Write an example of a goal within the context of workplace safety and an accompanying list of strategy points that can be used to accomplish the goal.
10. Why would reviewing incident reports be important when conducting an audit?
11. What safety committee activity would be beneficial to audit?

24 Security Auditing

Establishing a security audit will follow a similar process to that of establishing a fire or workplace safety audit. An evaluation of the facility will need to be performed to determine security risks that can result in loss. Though a uniform security audit can be created for an organization, it will be important to take into consideration the differences that exist between facilities.

Security audits are a customizable tool that can be effective in protecting both the organization and individual employees. These audits should be formatted with two issues in mind:

- Evaluate items that can allow outside influences to negatively impact employees
- Evaluate items that can provide the opportunity for internal loss to occur

Where fire and safety audits utilize standards and regulations, such as those promulgated by the Occupational Safety and Health Administration and the National Fire Protection Association, security auditing requires a more open approach due to there being a lack of specific regulatory guidance. Though some direction has been established for specific organizations that must comply with regulations promulgated by the Department of Homeland Security, security audits may vary greatly depending on the activity that occurs within a given organization or at a given facility. The first step in creating a security audit process is to identify what is hoped to be accomplished with the audit. This will require the use of an organizational evaluation to determine what areas of security will be addressed in the audit. Rather than simply depending on a body of regulations or standards to guide the process, risk must be evaluated to determine what processes can create the potential for loss to the organization and then an appropriate audit tool can be developed to address a system evaluation of security measures that should be in place.

SECURITY SURVEY

A starting point for a security auditing process can be the conduction of a security survey which may otherwise be known as a vulnerability assessment. A security survey is an open-ended approach to evaluating a facility or process that seeks to identify areas of weakness where loss can occur. Physical security can be addressed in a security survey by performing an inspection of the facility to determine if certain measures are in place. This might include the following:

- *Access Control*: Confirm that access to company property is monitored and screened by security personnel or technological tools, such as employee badge readers and automatic gates.

DOI: 10.1201/9781003371465-30

- *Lighting*: Determine if appropriate lighting is provided both in remote areas as well as those areas traveled by employees, contractors, and visitors.
- *Personnel Screening*: Observe individuals entering and exiting buildings to ensure that unauthorized access is prevented and evaluations of those departing buildings prevent theft from occurring through appropriate personal property inspections.
- *Fencing*: Evaluate the presence of fencing along the perimeter of the property to prevent unauthorized personnel from gaining access to the property.
- *Cameras*: Investigate the placement and coverage of cameras to determine if an appropriate level of monitoring can occur for sensitive areas throughout the facility and exterior property.

These and many other areas of vulnerability can be assessed using a security survey based on the details of the organization being evaluated. This survey can be seen as an audit through identifying weaknesses in security measures. The results of the security survey can also be utilized to build a systematic security audit tool that addresses the things that need to be in place to address loss.

FINANCIAL AUDITS

Whether there are cash issues as small as reconciling cash register sales at the end of a shift or a daily reconciliation of a petty cash box to something as large as electronically moving millions of dollars throughout an organization, financial auditing is a fundamental activity needed to prevent significant loss. Large-scale business failures have resulted in the Sarbanes-Oxley Act that has mandated financial stewardship and accountability to protect shareholder investment. It is imperative that the nature of the business is evaluated, applicable legal issues are identified, and a sound financial auditing system is implemented. Good people within an organization can make bad decisions when faced with financial opportunity for personal gain. Financial auditing systems will not only protect an organization against loss, but it will also communicate to employees that any theft of company assets will be quickly found.

Financial auditing may require partnerships to be established. Financial audits require a great deal of technical expertise to navigate profit and loss statements, balance sheets, and budget records to identify where loss is occurring. This may require a professional credential, such as a bachelor's degree in accounting or finance, being a Certified Public Accountant (CPA), and properly understanding accounting systems that are in use.

Once those with the appropriate degree of expertise are assembled an audit process can be established that will effectively evaluate the flow of money throughout an organization. Such an audit will also be able to ensure that appropriate controls are in place to prevent loss. For example, if a financial audit is being utilized to audit a dot-com business, one area of concern might be how returns are processed and credit given to a customer. The financial audit may be utilized to ensure controls are in place that will prevent an employee receiving the returned merchandise on the production floor from issuing a credit to a friend or family member. A friend may return a product that was purchased for $50, and the process may allow the employee to credit the friend with $500 on a credit card, thus allowing $450 in loss to occur to the organization. A financial audit would evaluate the weaknesses in such environments and allow for recommendations to secure the system.

Financial audits can also be utilized to evaluate the control that certain individuals have over fund or budget management. A member of management may have access to company finances that are redirected for personal gain. Financial auditing can be utilized to track the flow of money within the organization to identify the embezzlement of funds and to ensure that controls are in place that will prevent such activity from occurring.

On the less technical end of the spectrum, financial audits can be utilized to evaluate cash register or petty cash balances. Though such transactions may represent smaller individual points of loss, the risk can become great due to the volume of transactions that occur which could result in long-term systemic loss.

BEHAVIORAL AUDITS

Standard operating procedures should be in place to protect employees while they are at work. This would include such things as access control procedures. Behavioral audits can be established to observe the performance of these procedures to determine if they are executed properly and if enhancements need to be made to the system.

Behavioral audits are helpful in evaluating loss control procedures that are in place. A process may be in place for merchandise or product as it moves throughout a facility. Behavioral audits can be helpful in examining the performance of duties along the chain of events to ensure the work activity is being conducted as established.

The document utilized to conduct the behavioral audit should be designed to address the behaviors that are unique to the workplace and the designated procedures. Below is an example for evaluating access control:

Access Control	Yes	No	N/A
Employee, visitor, or contractor are properly greeted.		X	
Identification of entrant is checked.	X		
Visitor or contractor is signed in on the visitor Log.	X		
Visitor or contractor point of contact is notified.	X		
Visitor or contractor is placed in waiting area.	X		

Placing an "x" in either the "Yes" or "No" column provides the user with a scoring mechanism to provide metrics that can be monitored over time. Four "yes" responses and one "no" response equates to a score of 80%. This metric can be evaluated during future audits to determine if improvement is achieved or if regression occurs. Data showing this performance metric can be converted to a chart, such as the following, and communicated to applicable departments:

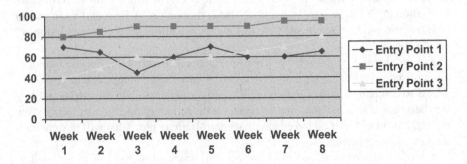

INVENTORY AUDITS

In all business environments some type of inventory is maintained. This is true whether it is a retail store, manufacturing plant, distribution center, or a governmental office. In a retail organization the inventory exists in the form of product on the shelf to be sold directly to a consumer or is stocked in a back storeroom. In manufacturing or distribution inventory exists in the form of what is in process of being shipped to an end-user or customer or material stored for production. In each of these situations, it is important to measure the amount of inventory loss that is occurring.

One inventory auditing tool is the cycle count. Inventory control departments typically exist in warehousing environments. Personnel working in inventory control will perform ongoing cycle counts to confirm the presence of inventory where it is supposed to exist. For example, smart phones stored in a warehouse may present a high risk for loss due to their value and ease of concealment. Cycle counts can be utilized to confirm the volume present as is indicated within the inventory system. A shipment of 2,000 smart phones may have been received on March 1. This shipment represents 20 different models which means each one will be designated a slot in warehouse racking. With the volume evenly distributed, there are 100 units of each model. Electronic records are maintained within the inventory control system that indicate the number ordered by customers and those that have been picked and shipped to fulfill the orders. Cycle counting is used to conduct interim counts to ensure that the correct number of each smart phone model is in stock. If a certain model number indicates shortage, theft may have occurred. This could prompt the review of security camera footage or the use of other investigative techniques to identify the source of the loss.

Not all loss identified through cycle counting can be considered theft. Inventory sampling conducted through the medium of cycle counting can be used to determine the volume of product loss. This volume of loss can be seen as shrink. Shrink is a metric that can be used to quantify product loss. Once this number is determined, the next step will be to assess what portion of that loss should be allocated to security-related loss. This activity might include the following process:

Item	Inventory Result
Merchandise in stock:	2,000 units (smart phones)
Inventory shortage (shrink):	30 units
• Damaged units found	5 units
• Units misplaced and recovered	10 units
Potential theft-related shrink	$30 - (5 + 10) = 15$

Once several items that can potentially be linked to theft-related shrink has been determined, it should be converted to a percentage of overall inventory. A percentage will provide a comparison between facilities of varying sizes or within the same facility throughout the year as inventory may fluctuate from month to month. The example above would provide a potential theft-related shrink metric of .75%. This number can be monitored each month to determine an increase or decrease.

GOALS AND STRATEGY

A structured set of security goals and corresponding strategy will help to keep an organization focused on appropriate activity. Security goals should be consistent in the two key areas of protecting people and protecting assets. As with other organizational goals, security goals must be high-level items that address significant challenges faced by the organization. Such goals could include:

- Reduce shrink by 20%
- Conduct security training among all current employees
- Improve access control behaviors by 30%

These goals are clearly measurable based on specified percentages of reduction or improvement and volume of training to be completed. The achievable aspect of goals must be carefully determined on a case-by-case basis. Setting a very bold goal to communicate a sense of urgency may discourage the group because they will immediately recognize the goal as unachievable. A more modest goal of reducing shrink by 20% might be more readily received than a 50% reduction goal. A level should be identified that will both indicate significant continuous improvement and will encourage the group to work toward it.

STRATEGY

Each security performance goal should contain a small number of strategy points that provide practical action to be taken that will influence goal attainment. Potential strategy points are listed below from the goals previously mentioned:

- Reduce shrink by 20%
  - Ensure applicable auditing techniques are in place and functioning
  - Install cameras to monitor high-risk inventory areas

- Train management on security issues that could affect their area of the facility
- Conduct security training among all current employees
  - Dedicate 1 hour each quarter to departmental training
  - Utilize technology and formats that are appealing to employees
- Improve access control behaviors by 30%
  - Ensure Security Officers have been properly trained in standard operating procedures
  - Provide immediate positive and negative feedback to Security Officers following behavioral observations

ACCOUNTABILITY

Accountability should be set among everyone at varying levels throughout the organization in order for security goals to be achieved. Accountability must be based on strategy activity, not on the goals themselves. Accountability must be set based on what strategic activity individuals can be engaged in that will influence accomplishing the goal. Activities to which individuals might be held accountable could include:

- Attend all appropriate security training
- Participate in a security audit of some type
- Provide a specified number of security recommendations for improvement

These activities are proactive and can influence the achievement of a goal. Each of these items is also within the control of an individual to accomplish.

Auditing security goals and strategy provide an opportunity to evaluate how applicable they are to the organization and to monitor progress toward achieving them. Feedback can be provided based on the evaluation to improve the effectiveness of this critical part of a security management system.

INCIDENT INVESTIGATION DATA

When the phrase "incident investigation" is used the image of an accident involving property damage or injury under the scope of workplace safety might come to mind. Though the area of safety may generate a wide range of incident investigations, security is also an area where auditing the quality of and data generated from incident investigations can be useful.

ALARM ACTIVATIONS

An investigation should be performed each time an alarm is activated. A weakness with security alarm systems is to become numb to their activation over time. They may be seen more as a nuisance than help to the organization. Consider auditing the following information when an alarm is activated, and a report is generated:

- Alarm point activated
- Shift working at the time of activation

- Date of activation
- Cause of activation
- Person responsible for activation

Collecting these data and placing them in graph form may alert the facility to issues to investigate. For example, when graphing information relating to the shift working at the time of alarm activation the following information might be found:

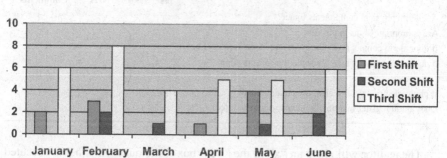

These data may lead to the need to investigate what things are occurring on third shift to cause such a difference in the volume of alarm activations. These data may also need to be tempered with information related to the cause of activations.

PRODUCT LOSS

An investigation should be conducted when product loss is identified. A thorough investigation will assist in determining if merchandise has been taken or if it has been misplaced due to a production error. If it is determined that a loss has occurred due to theft, every step necessary should be taken to ensure potential defects in the system have been identified and corrected to prevent future theft. Trending causes, times, and other variables related to theft can be utilized over time to determine larger systemic issues that need to be addressed within the organization.

An audit is an opportunity to review how well a facility is doing in the management of this process. Auditing incident investigations will help to establish a perspective of how well the facility is investigating incidents and following up on corrective action to ensure future incidents do not occur.

FACILITY INSPECTIONS

A physical inspection of the facility could be established on a periodic basis to ensure all physical security systems are in place and properly functioning. As with safety inspections, subjectivity can be barrier to conducting and obtaining usable information from an inspection of the facility. Subjectivity can be controlled to a great degree through:

- Clearly defining what "good" or "acceptable" looks like
- Calibrating auditors
- Establishing consistent line items to be rated

A section from a security inspection could be as follows:

PERIMETER ALARM TESTS

	Yes	No	N/A	Comments
Are alarm physical components intact?				
Are appropriate signs in place?				
If door alarm contact is in place, does the door open readily?				
If door alarm contact is in place, do the sensors align properly?				
If photo-electric eye, is a clear area maintained?				
Does alarm register in a timely manner once activated?				

The auditor will place an "x" in the "Yes" box if the question is being executed in an acceptable manner and will place an "x" in the "No" box if the line item is found to be unsatisfactory. "N/A" will be marked with an "x" if that line item does not apply to the facility being audited. A spreadsheet can sum all the boxes marked with an "x" to determine the total possible points for the facility and then calculate a percentage of "yes" responses for a score.

It is interesting to note that though this audit may be construed as ensuring merchandise stays inside the facility, it can also be utilized to ensure the safety of employees during an emergency. In this example of perimeter doors, it is important to secure them in a way that protects product, but also allows employees to safely exit in the event of an emergency. This would exclude the use of items such as chains and padlocks to secure doors intended for employees to use in the event of a fire. Panic hardware and effective alarm sensors will allow employees to exit and provide management with the opportunity to respond when the alarm is activated.

CASE STUDIES

FIRE

Jessie has been with the fire department for 15 years and there has always been a great deal of trust among all the firefighters. The last two years has seen an unusual amount of turnover causing several new people to enter the department. A few of the veterans have reported a couple of missing tools and pieces of equipment. The last petty cash count was a little short, but Jessie assumed there were just a couple of outstanding receipts.

SAFETY

Seth has just been placed in charge of his first capital project as the site Safety Manager. He is jointly managing a $150,000 fall protection construction project with the site Maintenance Manager. A month into the project, Seth was conducting a review of the work completed and money spent so he could report project progress to the Plant Manager. He noticed a $5,000 discrepancy. $85,000 worth of work has been completed and billed, but the account appears to have released $90,000.

SECURITY

Sally is the Security Manager of a distribution center that receives, stores, and ships electronic merchandise. She has partnered with the Inventory Control Manager to monitor cycle count results in some of the high-risk product areas. She has noted a gradual increase in the percentage of potentially theft-related loss.

- How might a security audit assist the evaluation of the situation?
- What areas of financial or general security auditing might be of concern?

EXERCISES

For the following questions, identify a single facility environment in which you would like to situate your responses and answer each question accordingly. Answer each question in the context of a fire, safety, or security audit.

1. What issues would be necessary to include and evaluate when conducting a security survey?
2. What role would financial auditing play in the organization? What financial processes would you audit? What expertise would be needed?
3. How could cycle counting help to identify loss? Is all loss identified in cycle counting automatically considered as theft-related? Why or why not?
4. What is "shrink"? How can you quantify the amount of shrink that can be considered as potentially theft-related?
5. Why might the review of goals and strategy be helpful as a component of the audit process?
6. What types of investigations might yield useful information in the audit process?
7. How is a physical inspection useful when conducting an audit?

25 Fire Auditing

Three primary areas of auditing are examined in this text, which are fire, safety, and security. The area of fire auditing has two applications:

- Industrial identification of fire risks and including applicable information in basic and complex audits
- Fire service auditing

INDUSTRY

Fire risks can be evaluated, and applicable standards or regulations can be used to develop basic audits that can be used to address exposures on a daily, weekly, monthly, or quarterly basis. Fire issues can also be incorporated into a complex audit used to evaluate a given facility on an annual basis. This can occur through such applications as:

- *Daily Audits*: Facilities that must address the risk of dust explosions may perform daily documented audits of work areas where dust can collect.
- *Weekly Audits*: A distribution center may perform weekly departmental audits of fuel load management, such as storage of flammable materials or the storage of merchandise in proximity to fire system sprinklers.
- *Monthly Audit*: A manufacturing plant may include a review of fire hazards in a monthly safety audit.
- *Quarterly Audit*: A food processing facility might conduct quarterly audits of fire suppression systems to ensure they operate properly, and qualified workers or managers exist among the maintenance or safety staff to ensure their operation during an emergency.
- *Annual Audit*: Fire hazards might be included in an annual audit that is conducted across a corporation.

Each of these applications can address the assessment of how well fire risks are being managed on various levels. Daily audits are conducted on a granular level to address specific high-risk exposures, where annual audits address the management of fire hazards from a system management perspective.

Standards and regulations must be reviewed to determine what requirements apply to establish an audit document. The two primary sources for this information are the Occupational Safety and Health Administration (OSHA) and the National Fire Protection Association (NFPA). OSHA promulgates workplace safety regulations with which employers must comply. The scope or applicability sections within each regulation will help each place of employment determine how or if a

DOI: 10.1201/9781003371465-31

given regulation applies to their environment. Regulations that address fire hazards include:

GENERAL INDUSTRY

- 1910.34 – coverage and definitions for means of egress
- 1910.36 – design and construction of exit routes
- 1910.37 – maintenance, safeguards, and operational features of exit routes
- 1910.38 – emergency action plans
- 1910.39 – fire prevention plans
- 1910.155 – scope, application, and definitions applicable to fire protection regulations
- 1910.156 – fire brigades
- 1910.157 – portable fire extinguishers
- 1910.158 – standpipe and hose systems
- 1910.159 – automatic sprinkler systems
- 1910.160 – fixed extinguishing system general requirements
- 1910.161 – fixed extinguishing systems (dry chemical)
- 1910.162 – fixed extinguishing systems (gaseous agent)
- 1910.163 – fixed extinguishing systems (water spray and foam)
- 1910.164 – fire detection systems
- 1910.165 – employee alarm systems
- 1910.251 – welding, cutting, and brazing definitions
- 1910.252 – welding, cutting, and brazing general requirements
- 1910.253 – oxygen-fuel gas welding and cutting
- 1910.254 – arc welding and cutting
- 1910.255 – resistance welding

CONSTRUCTION

- 1926.24 – fire protection and prevention
- 1926.34 – means of egress
- 1926.35 – employee emergency action plans
- 1926.150 – fire protection
- 1926.151 – fire prevention
- 1926.152 – flammable and combustible liquids
- 1926.153 – liquefied petroleum gas
- 1926.154 – temporary heating devices
- 1926.155 – fire protection and prevention definitions
- 1926.156 – fixed extinguishing system general requirements
- 1926. 157 – fixed extinguishing systems (gaseous agent)
- 1926.158 – fire detection systems
- 1926.169 – alarm systems
- 1926.352 – fire prevention (welding, cutting, and heating)

- 1926.353 – ventilation and protection in welding, cutting, and heating
- 1926.354 – welding, cutting, and heating in way of preservative coatings
- 1926 Subpart U – blasting and the use of explosives

SHIPYARDS

- 1915.501 – fire protection general provisions
- 1915.502 – fire safety plan
- 1915.503 – precautions for hot work
- 1915.504 – fire watches
- 1915.505 – fire response
- 1915.506 – hazards of fixed extinguishing systems on board vessels and vessel sections
- 1915.507 – land-side fire protection systems
- 1915.508 – training
- 1915.509 – definitions applicable to fire protection regulations

MARINE TERMINAL

- 1917.21 – open fires
- 1917.30 – emergency action plans
- 1917.122 – employee exits
- 1917.152 – welding, cutting, and heating

LONGSHORING

- 1918.100 – emergency action plans

This list of regulations directly impacts the management of fire hazards in the workplace. In addition, fire hazards may be specifically addressed within the scope of other regulations. For example, OSHA's grain handling standard (1910.272) addresses several hazards in grain elevators that could result in a fire or explosion. OSHA regulations would need to be reviewed and properly applied to the given place of employment.

FIRE SERVICE

The fire service provides a unique application for the practice of auditing. In addition to fire auditing, this text also addresses safety and security auditing. The concept of fire auditing as applied specifically to the fire service can be accomplished by addressing standard facility safety and security issues, which are discussed in other chapters in this text. Fundamental issues of safety auditing can apply by auditing a given fire station to determine if all measures are in place to protect firefighters. Security auditing can be conducted to determine if the facility and assets are adequately protected.

Fire service auditing can also be accomplished by exploring NFPA standards and develop both basic and complex audits to address issues as needed. Basic audits can be developed to perform routine evaluations of certain risks. For example, fire apparatus can be audited on a routine basis to ensure that all systems are properly functioning, and that all equipment is available. A complex audit can be performed to address a broad spectrum of issues based on the desired scope of the audit. Both OSHA regulations and NFPA standards can apply to a given fire department. OSHA information is addressed in the chapter that reviews safety auditing, so the focus here will be on exploring NFPA standards that are applicable to a fire department. Below are categories of NFPA standards that could be explored and used to develop both basic and complex audits.

EQUIPMENT

- NFPA 10 – Standard for Portable Fire Extinguishers
- NFPA 412 – Standard for Evaluating Aircraft Rescue and Fire-Fighting Foam Equipment
- NFPA 1801 – Standard on Thermal Imagers for the Fire Service
- NFPA 1932 – Standard on Use, Maintenance, and Service Testing of In-Service Fire Department Ground Ladders
- NFPA 1936 – Standard on Powered Rescue Tools
- NFPA 1961 – Standard on Fire Hose
- NFPA 1962 – Standard for the Inspection, Care, and Use of Fire Hose, Couplings, and Nozzles and the Service Testing of Fire Hose
- NFPA 1963 – Standard for Fire Hose Connections
- NFPA 1964 – Standard for Spray Nozzles
- NFPA 1965 – Standard for Fire Hose Appliances

APPARATUS

- NFPA 414 – Standard for Aircraft Rescue and Fire-Fighting Vehicles
- NFPA 1901 – Standard for Automotive Fire Apparatus
- NFPA 1906 – Standard for Wildland Fire Apparatus
- NFPA 1911 – Standard for the Inspection, Maintenance, Testing, and Retirement of In-Service Automotive Fire Apparatus
- NFPA 1912 – Standard for Fire Apparatus Refurbishing
- NFPA 1925 – Standard on Marine Fire-Fighting Vessels

PERSONAL PROTECTIVE EQUIPMENT

- NFPA 1851 – Standard on Selection, Care, and Maintenance of Protective Ensembles for Structural Fire Fighting and Proximity Fire Fighting
- NFPA 1852 – Standard on Selection, Care, and Maintenance of Open-Circuit Self-Contained Breathing Apparatus (SCBA)
- NFPA 1951 – Standard on Protective Ensembles for Technical Rescue Incidents

- NFPA 1952 – Standard on Surface Water Operations Protective Clothing and Equipment
- NFPA 1971 – Standard on Protective Ensembles for Structural Fire Fighting and Proximity Fire Fighting
- NFPA 1975 – Standard on Station/Work Uniforms for Emergency Services
- NFPA 1977 – Standard on Protective Clothing and Equipment for Wildland Fire Fighting
- NFPA 1981 – Standard on Open-Circuit Self-Contained Breathing Apparatus (SCBA) for Emergency Services
- NFPA 1982 – Standard on Personal Alert Safety Systems (PASS)
- NFPA 1983 – Standard on Life Safety Rope and Equipment for Emergency Services
- NFPA 1989 – Standard on Breathing Air Quality for Emergency Services Respiratory Protection
- NFPA 1991 – Standard on Vapor-Protective Ensembles for Hazardous Materials Emergencies
- NFPA 1992 – Standard on Liquid Splash-Protective Ensembles and Clothing for Hazardous Materials Emergencies
- NFPA 1994 – Standard on Protective Ensembles for First Responders to CBRN Terrorism Incidents
- NFPA 1999 – Standard on Protective Clothing for Emergency Medical Operations

QUALIFICATIONS

- NFPA 405 – Standard for the Recurring Proficiency of Airport Fire Fighters
- NFPA 472 – Standard for Competence of Responders to Hazardous Materials/Weapons of Mass Destruction Incidents
- NFPA 473 – Standard for Competencies for EMS Personnel Responding to Hazardous Materials/Weapons of Mass Destruction Incidents
- NFPA 1000 – Standard for Fire Service Professional Qualifications Accreditation and Certification Systems
- NFPA 1001 – Standard for Fire Fighter Professional Qualifications
- NFPA 1002 – Standard for Fire Apparatus Driver/Operator Professional Qualifications
- NFPA 1003 – Standard for Airport Fire Fighter Professional Qualifications
- NFPA 1005 – Standard for Professional Qualifications for Marine Fire Fighting for Land-Based Fire Fighters
- NFPA 1006 – Standard for Technical Rescuer Professional Qualifications
- NFPA 1021 – Standard for Fire Officer Professional Qualifications
- NFPA 1026 – Standard for Incident Management Personnel Professional Qualifications
- NFPA 1031 – Standard for Professional Qualifications for Fire Inspector and Plan Examiner
- NFPA 1033 – Standard for Professional Qualifications for Fire Investigator
- NFPA 1035 – Standard for Professional Qualifications for Public Fire and Life Safety Educator

- NFPA 1037 – Standard for Professional Qualifications for Fire Marshal
- NFPA 1041 – Standard for Fire Service Instructor Professional Qualifications
- NFPA 1051 – Standard for Wildland Fire Fighter Professional Qualifications
- NFPA 1061 – Standard for Professional Qualifications for Public Safety Telecommunicator
- NFPA 1071 – Standard for Emergency Vehicle Technician Professional Qualifications
- NFPA 1081 – Standard for Industrial Fire Brigade Member Professional Qualifications

TRAINING

- NFPA 1401 – Recommended Practice for Fire Service Training Reports and Records
- NFPA 1402 – Guide to Building Fire Service Training Centers
- NFPA 1403 – Standard on Live Fire Training Evolutions
- NFPA 1404 – Standard for Fire Service Respiratory Protection Training
- NFPA 1407 – Standard for Training Fire Service Rapid Intervention Crews
- NFPA 1410 – Standard on Training for Initial Emergency Scene Operations
- NFPA 1451 – Standard for a Fire Service Vehicle Operations Training Program
- NFPA 1452 – Guide for Training Fire Service Personnel to Conduct Dwelling Fire Safety Surveys

OPERATIONS

- NFPA 13E – Recommended Practice for Fire Department Operations in Properties Protected by Sprinkler and Standpipe Systems
- NFPA 403 – Standard for Aircraft Rescue and Fire-Fighting Services at Airports
- NFPA 450 – Guide for Emergency Medical Services and Systems
- NFPA 1201 – Standard for Providing Emergency Services to the Public
- NFPA 1250 – Recommended Practice in Emergency Service Organization Risk Management
- NFPA 1405 – Guide for Land-Based Fire Fighters Who Respond to Marine Vessel Fires
- NFPA 1620 – Standard for Pre-Incident Planning
- NFPA 1670 – Standard on Operations and Training for Technical Search and Rescue Incidents
- NFPA 1710 – Standard for the Organization and Deployment of Fire Suppression Operations, Emergency Medical Operations, and Special Operations to the Public by Career Fire Departments
- NFPA 1720 – Standard for the Organization and Deployment of Fire Suppression Operations, Emergency Medical Operations and Special Operations to the Public by Volunteer Fire Departments

INDUSTRIAL

- NFPA 61 – Standard for the Prevention of Fires and Dust Explosions in Agricultural and Food Processing Facilities
- NFPA 69 – Standard on Explosion Prevention Systems
- NFPA 70 – National Electrical Code
- NFPA 70B – Recommended Practice for Electrical Equipment Maintenance
- NFPA 70E – Standard for Electrical Safety in the Workplace
- NFPA 101 – Life Safety Code
- NFPA 101A – Guide on Alternative Approaches to Life Safety
- NFPA 600 – Standard on Industrial Fire Brigades

SECURITY

- NFPA 601 – Standard for Security Services in Fire Loss Prevention
- NFPA 730 – Guide for Premises Security
- NFPA 731 – Standard for the Installation of Electronic Premises Security Systems

SAFETY AND HEALTH

- NFPA 1500 – Standard on Fire Department Occupational Safety and Health Program
- NFPA 1521 – Standard for Fire Department Safety Officer
- NFPA 1561 – Standard on Emergency Services Incident Management System
- NFPA 1581 – Standard on Fire Department Infection Control Program
- NFPA 1582 – Standard on Comprehensive Occupational Medical Program for Fire Departments
- NFPA 1583 – Standard on Health-Related Fitness Programs for Fire Department Members
- NFPA 1584 – Standard on the Rehabilitation Process for Members During Emergency Operations and Training Exercises

The standards listed here can be used to form an audit document for a fire department to determine how well it is performing according to established standards published by the NFPA. A basic audit might address only one topic, whereas a complex annual audit might take a large spectrum of issues into consideration.

CASE STUDIES

FIRE

Chad has been tasked with creating an audit program for the Sunnyville Fire Department. SFD has experienced a sharp increase in firefighter injuries over the past 2 years. One avenue Chad could pursue is creating an audit that focuses on the conditions of the station and emergency scene safety issues. A national fire conference is in the process of accepting proposals from potential speakers and Chad is also aware that a presentation on a fire station audit based on NFPA would be a unique topic and would be beneficial for others. He begins to think about how he should structure his audit.

SAFETY

A production line at Acme Manufacturing continues to experience periodic small fires due to the presence of combustible dust. Whitney, the facility Safety Manager has worked with the department through employee training and fire prevention protocols, but the problem continues. She wants to create a systematic audit process to track the conditions in the department to help ensure that fire hazards are controlled.

SECURITY

Shawn is the Regional Loss Prevention Manager for Acme Auto Parts. He is tasked with managing safety issues in addition to security. Acme has a robust security audit tool, but nothing is in place that clearly addresses safety. One problem that Shawn has noticed in several stores is that product is stored too close to sprinkler heads. He recognizes that the sprinkler system would not function properly if a fire occurred due to being obstructed by the stored boxes of merchandise. He has mentioned this to a number of store managers but is met with the common response that they must store products in that fashion due to the size of the stock rooms. Shawn would like to address this issue formally through an audit mechanism that could be used to communicate the issue to upper management on an ongoing basis.

- What variables should the auditor take into consideration in deciding which direction to take the audit?
- What should be the scope of the audit?

EXERCISES

For the following questions, identify a single facility environment in which you would like to situate your responses and answer each question accordingly. Answer each question in the context of a fire, safety, or security audit.

1. How can fire audits be applied in the general industry?
2. How can fire auditing be applied to both basic and complex audits?
3. What two organizations publish information that can be useful in creating a fire audit?
4. What is unique about the application of auditing to fire departments?
5. What categories of NFPA standards exist that could be used to build an audit program for a fire department?

26 Creating Mini Audits

A completed management system audit document holds the potential and agility to be used to create risk-based mini audits. Pieces of the audit document can be extracted and used to rapidly collect data that can be used to address current and emerging issues in the workplace. The risk-based mini audit can utilize one or all the sections of documentation review, facility inspection, and worker interviews, depending on the nature of the issues that are being experienced within the organization. Topical mini audits can also be developed and conducted at the same time of the management system audit when there is a lack of auditor resources or time constraints dictate the necessity to combine audit processes.

RISK-BASED MINI AUDITS

A risk-based mini audit can be created by dissecting the necessary components of a management system audit. The appropriate pieces can be assembled to assess the behavior or property where the risk has presented itself. For example, maintenance technicians and contractors might experience several fires while cutting and welding repair has been occurring on the production floor. Cutting and welding, otherwise known as "hotwork," is addressed in the safety management system audit, which includes the following aspects of assessment:

- *Documentation Review*: a review is performed that addresses the following records:
  - Written Hotwork Program
  - Training records
  - Hotwork permits that have been completed throughout the year
- *Facility Inspection*: the following aspects are observed:
  - Condition of hotwork equipment
  - Condition of PPE used when hotwork is being performed
  - Condition of equipment used to prevent and respond to fires
  - Performance of hotwork on the production floor
- *Worker Interviews*: knowledge of workers is explored to include:
  - How to obtain a hotwork permit
  - Proper procedures that must be followed when performing hotwork
  - PPE that must be worn when performing hotwork
  - Emergency response procedures in the event of a fire

It will be appropriate to examine this scope of items and activities as a part of a safety management system audit. However, only certain items might be of use to create a risk-based mini audit to collect data quickly to determine why fires are occurring. Items that have a direct impact on the occurrence of fires should be

DOI: 10.1201/9781003371465-32

included. For example, a review of all documents might not be necessary. A review of the program might not be helpful due to an assessment of the program occurring during the safety management system audit with appropriate improvements being addressed. A review of the following documents might prove helpful to rapidly collect useful data:

- Training records to ensure all current maintenance technicians have received training
- Contractor management documentation from the Contractor Management Program that indicates all current contractors have been informed of proper hotwork procedures
- Fire incident investigation documents from the Incident Investigation Program to determine trends that might be present
- Analysis of hotwork permits to ensure they have been properly completed
- Observations of hotwork conducted to determine potential issues

The Hotwork Program facility inspection component of the safety management system will be of use to closely examine hotwork activities each time they are performed. Frequent observations performed to collect data on safe and at-risk behavior in each step of the task can be used to identify potential work process issues that could be causing the fires.

In-depth employee interviews can be conducted that begin with questions utilized in the safety management system audit. Further detail can be explored by interviewing workers about specific things that occur throughout the process of performing hotwork. Interviews can result in new perspectives being brought to the problem and additional information that can be used to identify opportunities for improvement.

Some parts of the risk-based mini audit will be conducted on a one-time basis while others will be conducted numerous times to collect data. For example, a review of training records and conducting worker interviews would be a one-time exercise. Behavioral observations of hotwork being performed would occur multiple times whenever hotwork occurs.

TOPICAL MINI AUDITS

Available resources to conduct audits might vary between management systems. Occupational safety management systems have evolved greatly to include employers investing in hiring safety professionals at various levels in an organization. Such human resources investment has provided a robust opportunity to carry out safety management system audits. The realm of security management might not be as fortunate. In some situations, a security manager might exist with no additional staffing. In other cases, security might be an additional responsibility of someone in management, such as the safety manager. Security management system audits will need to be conducted regardless of the level of human resources allocated to the discipline.

Topical mini audits can be developed to address the critical few issues that affect the performance of the management system. These can be brief documents that can be reviewed in a short amount of time in situations where human resources are not in place to conduct in-depth management system audits. An organization might be well-staffed in relation to the occupational safety management system with only on security manager. Two options are available:

- A security mini audit can be developed that addresses critical management system issues and be conducted by the security manager. The security manager can organize a process through which the mini audit can be performed at all facilities.
- If the security manager cannot feasibly carry out the mini audit at all the appropriate facilities, occupational safety management system auditors can be trained on how to perform the security mini audit and conduct it as an additional task when they perform safety management system audits. Collaboration can occur between the security manager and occupational safety manager to meet the needs of security management system assessment.

Flexibility can be used to adapt the use of topical mini audits to meet the need of management system assessment. A review of human resources can reveal where needs exist to deliver topical mini audits to ensure all management systems receive the attention and assessment needed to ensure continuous improvement.

CASE STUDIES

FIRE

Susan is responsible for fire prevention and education in her city. She has reviewed incident reports from the past calendar year and determined that there is an increase in industrial fires from prior years that are associated with contractor activities on the sites of manufacturing plants and distribution centers. She sees the need to further explore the problem but has limited resources to accomplish the task.

SAFETY

Cliff is the Corporate Safety Director for a manufacturing company where he has implemented a comprehensive safety management system auditing process. A recent spike in ergonomic incidents has had a great impact on incident rate data as well as workers' compensation costs. He sees the need to further explore ergonomic program implementation. Cliff has also been approached by the Corporate Security Manager who is impressed with the safety management system auditing process that has been implemented. She would like to do something similar but does not have the personnel available that Cliff has in the form of regional and site occupational safety professionals.

SECURITY

Farlon is the Security Manager for a grocery retail chain. He has worked to develop a security management system and has expanded it to include an annual comprehensive security management system audit to determine where opportunities for improvement exist at each location. In a recent analysis of management system incident reports, he noted an increase in workplace violence incidents in urban facilities.

- How can a risk-based mini audit be applied to improve management system performance?
- How can a topical mini audit be applied to improve management system performance?

EXERCISES

For the following questions, identify a single facility environment in which you would like to situate your responses and answer each question accordingly. Answer each question in the context of a fire, safety, or security audit.

1. How can documentation review questions be applied when conducting risk-based mini audits?
2. How can facility inspection questions be applied when conducting risk-based mini audits?
3. How can employee interview questions be applied when conducting risk-based mini audits?
4. What issues of flexibility might need to be applied when conducting assessments on a one-time basis?
5. What issues of flexibility might need to be applied when conducting assessments on an ongoing basis to collect a larger volume of data to address an emerging issue?
6. How can topical audits be implemented by engaging teammates from multiple management systems?
7. How can a topical audit be implemented by a subject matter expert alone when there is no access to other human resources?

Part VII

Audit Program Opportunities

27 The Audit as a Training Tool

Though the activity of conducting an audit is primarily designed to measure the performance of a facility in relation to issues that are within the scope of the audit, an audit can also be used as a training tool. The audit content can be used both as a measurement of performance and an opportunity to train facility personnel on ways in which compliance can be achieved. Text bubbles that provide instructive information were discussed in the chapter that discussed the construction of the audit document. Utilizing the audit as a training tool is an expansion of the concept of using such information within the process of conducting an audit. This can be accomplished by implementing the following:

- Identify who is a potential target for training
- Take advantage of teachable moments
- Exercise good interpersonal communication skills
- Train in the classroom and on the floor

POTENTIAL TARGETS FOR TRAINING

The auditor will encounter a number of individuals throughout the process of conducting an audit. This might include the following individuals depending on the type and environment of the audit:

- Plant Manager
- Safety Manager
- Security Manager
- Shift Managers
- Human Resources Manager
- Maintenance Manager
- Procurement Manager
- Finance Manager
- Department Supervisors
- Employees
- Contractors
- Fire Chief
- Fire Officers
- Training Officer
- Firefighters

DOI: 10.1201/9781003371465-34

Due to the length of this list, the auditor will need to exercise discretion as to which individuals should be targeted for training. Training cannot be effectively conducted among such a large population if everyone is a target for consideration. The auditor will need to identify a small list of people who could benefit from such an opportunity. The auditor can begin with identifying the individual who is primarily responsible for the material covered within the scope of the audit. This might include:

- Safety Audit – Safety Manager
- Security Audit – Security Manager
- Fire Audit – Safety Manager; Fire Department Training or Safety Officer

Once the person who is primarily responsible for the topic covered by the audit has been identified, the auditor can then prioritize a small list of a few individuals who are most closely connected with issues that impact the success of items covered within the scope of the audit. This might include:

- Maintenance Manager responsible for the implementation of many programs and projects that are covered by the audit
- Shift Managers who direct the activity of department supervisors, which can set the tone for organizational culture activity, such as discussing appropriate topics in various meetings
- Finance Manager who influences budget decisions made within the organization that can impact activity within the scope of the audit

These individuals are not ultimately responsible for the discipline covered by the audit, but they are instrumental in the success of program implementation. Individuals such as the Plant Manager and Fire Chief are not on this list. Though these individuals are the primary leaders at the facility being audited, they may not typically be highly engaged in the audit process, thus limiting the opportunity for an auditor to utilize training as a tool when interacting with them. Individuals in high leadership positions will typically only be provided direct communication regarding the progress of the audit and audit results. An exception to this is if the facility is very small in size. For example, a Plant Manager of a small business that employs 20 people would be a primary target for training due to such an individual being both the primary leader and having direct responsibility for managing the area that is covered by the audit.

TEACHABLE MOMENTS

Training during an audit is an informal experience. There are no outlines, PowerPoint presentations, or structured exercises. It is the opportunity to take advantage of situations as they arise to train someone on a way to better accomplish a task that is associated with the audit. Such situations are considered teachable moments in that they are informal opportunities that spontaneously arise that provide an opportunity for the auditor to train an individual on a certain issue.

Informal learning can occur throughout the audit process. The auditor can take advantage of opportunities as they arise to provide informal training. For example, the auditor and Maintenance Manager could be conducting the plant inspection together. They observe work being done overhead and the auditor notes that a maintenance technician is wearing a harness and lanyard, but the lanyard is tied off to a catwalk handrail. The auditor immediately recognizes the problem due to a fall protection anchorage point needing to support 5,000 pounds of force, but a handrail only needing to withstand a load of 200 pounds. The auditor asks that the work be immediately stopped due to the threat posed to the maintenance technician. She then takes the opportunity to use this incident as a teachable moment. She talks to the Maintenance Manager about reasons why workers might tie off to a handrail and asks for suggestions that might address the problem. The interaction serves as an opportunity for the Maintenance Manager to gain a greater understanding of fall protection and unique applications that can ensure the protection of workers.

INTERPERSONAL COMMUNICATION SKILLS

The auditor must exercise exceptional interpersonal communication skills for teachable moments to be effective. Though the auditor may be perceived as being the subject matter expert in the area that is covered by the audit, there needs to be a sense of equality in the interaction between the auditor and those being audited. The auditor can accomplish this by exercising several skills:

- *Listen Before Speaking*: The auditor should listen intently to what is being said by those in the audit environment to gain an understanding of their perspective and level of technical knowledge. The words that are spoken should be taken into context by understanding the background and influences that have caused the comments to be made. Absorbing and understanding this information will help the auditor to make informed and courteous responses.
- *Avoid Condescension*: The auditor should be the subject matter expert among those present during an audit. This includes being an expert on all organizational policies and regulatory issues covered within the scope of the audit as well as how they should be implemented in the environment. However, this expertise should not result in a condescending tone when speaking with those who are present during an audit. Condescension in communication can easily result in a negative response from those at the facility. The primary tone in the conversation should be one of mutual respect and teamwork in evaluating and identifying solutions to issues.
- *Be Direct*: Be clear when addressing issues. Provide specific examples that are evidence of issues identified during the audit. Being vague can result in confusion and frustration on the part of facility personnel. For example, if it is believed that a confined space permit should have been issued during a certain entry, be direct and specific as to what hazard existed that would require the use of the permit and subsequent precautions to be taken.

Utilizing interpersonal communication skills will help to foster a positive audit experience and can influence the degree to which location personnel are open to communication that fosters the use of the audit as a training tool. Though auditors are the subject matter experts on issues covered within the scope of the audit, they are not necessarily the subject matter experts on facility operations. This sets the stage for the auditor to utilize interpersonal communication skills to teach issues related to the audit as well as learn how implementation might be impacted by what occurs at the facility.

TRAINING ENVIRONMENT

The traditional training environment might be considered the classroom where the trainer imparts knowledge to trainees or a production floor where trainees engage in exercises. The training environment during an audit is much more dynamic. The auditor can take advantage of several environments to conduct training as the schedule used to conduct the audit unfolds:

- *Office*: A member of facility management should be with, or at least nearby, the auditor while the documentation review occurs. As the auditor identifies issues or has questions, the member of facility management can immediately interact with the auditor. This provides an opportunity for training to occur in that the auditor can provide guidance on ways in which the documentation can be corrected, or a system put in place to better manage it.
- *Work Areas*: The facility inspection will require the use of at least one member of management to serve as a guide to navigate the facility. Training can occur throughout the facility inspection by discussing opportunities for improvement and how the improvement can be achieved.
- *Breaks*: Whether it is a short break or an hour eating lunch, casual periods of time present an opportunity to train location personnel. These informal times provide a relaxed environment in which issues can be discussed and instruction provided by the auditor to address certain challenges.
- *Closing Conference*: The closing conference can be used to address significant potentials for management system improvement. The auditor can identify major issues that can be addressed to impact management system improvement and provide guidance on potential solutions to challenges.

CASE STUDIES

FIRE

Robert is in the process of conducting the annual safety audit at Station 1 when he realizes that an electrical closet containing the breaker panels for the station is being used as a storage area. Boxes of supplies and unused equipment are stacked in the closet making it difficult to access electrical panels. He knows that three feet must be maintained in front of each panel to ensure quick access in the event of an emergency. Mary, a Station Captain who has been accompanying Robert on the facility inspection, states that this has been the case for the area for as long as she has worked at the station.

SAFETY

Patricia is working her way through the documentation review when she realizes that only one-third of the hotwork permits are properly completed. Errors appear to be in numerous sections of different permits, which provide a lack of a theme in the errors. She asks Bob, the Maintenance Manager, about the hotwork permit process that is in place at the facility.

SECURITY

Andrea is performing a facility inspection when she notices an exit door that is blocked open. She performs a closer inspection of the door and finds that there is a door contact with a local alarm, but the local alarm has been disconnected. She asks Evan, the facility Security Manager who has been accompanying her on the inspection, why the door is blocked open. He responds by saying that area of the facility receives poor ventilation, so the door is propped open to provide a flow of air for the employees working in a nearby department.

- How can training be used in these situations as a tool to enhance the facility's performance?
- What communication skills might be utilized to ensure a favorable outcome of the interaction?

EXERCISES

For the following questions, identify a single facility environment in which you would like to situate your responses and answer each question accordingly. Answer each question in the context of a fire, safety, or security audit.

1. Who might be a target for training during an audit? Why would those individuals be selected?
2. How can you take advantage of a teachable moment during an audit?
3. In what ways can informal learning occur during an audit?
4. How can you exercise interpersonal communication skills to help ensure a favorable outcome of an interaction with a member of location management?
5. What environments might be used to conduct training as an audit unfolds?

28 Employee Involvement

Opportunities exist within the process of conducting an audit to involve employees. An audit may typically be viewed as an event that takes place among members of management, but employees can become engaged in different aspects of an audit. Following are avenues through which employees can be engaged while an audit is conducted at their place of employment:

- Employee recommendations
- Safety Committee activity prior to the audit
- Assigned programs for the documentation review portion of the audit
- Act as the escort for the facility inspection

OSHA supports the inclusion of employees as an integral part of a well-functioning safety and health program. Employee engagement should not be a superficial effort to make them feel part of the organization but should be an authentic activity that demonstrates the value that their participation brings to the success of the organization. Their involvement can be critical in the success of the audit and the implementation of the management system.

EMPLOYEE RECOMMENDATIONS

Employees can make recommendations regarding issues that affect performance on the audit. Employees work in the departments and hold jobs where implementation of programs occurs, so it is they who can have some of the greatest insight as to how success can be most effectively achieved. For example, if the audit has historically shown deficiencies in confined space entry, employees could make recommendations for addressing the issues that have been identified.

Management must support employees in making recommendations and must create an environment in which this can occur. Employees need to feel safe to make recommendations regarding program improvements. Rather than simply accept the Confined Space Program as it was delivered to them, they may know how to better manage the process in a way that both ensures compliance and promotes employee safety. However, such recommendations may not be made if there is an environment where they fear repercussions for making corrective comments regarding the program. They may fear that management will perceive this as criticizing the work that has been done to create the program. Management can resolve this potential barrier by openly communicating that employee suggestions are welcome.

SAFETY COMMITTEE

The Safety Committee, or similar organization, has been a prominent organization through which employees can become involved in fire, safety, and security programs. Beyond simply making recommendations, the Safety Committee is often charged

DOI: 10.1201/9781003371465-35

with directly working to implement solutions and engaging with management on issues that are beyond their ability to address. Safety Committee members go beyond making recommendations for improvement by also recommending solutions for the improvement. For example, the facility might be vulnerable to workplace violence due to deficiencies in the access control system. Rather than drawing attention to the issue and expecting management to solve the problem, Safety Committee members can engage in the activity of problem-solving by generating potential solutions. Of the solutions that have been generated, the Safety Committee can then decide on the ones that they believe will have the highest degree of likelihood for success and present them to management. The Safety Committee can then follow up with management on what final decisions have been made.

Safety Committee membership can involve several roles. These might include:

- Chairperson
- Vice Chairperson
- Secretary
- Subcommittee Chairperson
- Member

Each of these roles should be accompanied by defined responsibilities. Safety Committee members will not come to meetings and sit idly but will engage in the process by actively participating through executing their responsibilities.

DOCUMENTATION REVIEW

Employees can be placed in charge of walking the auditor through certain programs for which they are assigned. Rather than a member of management standing by as a point of contact for the auditor, various employees can be scheduled to stand by to answer questions regarding given programs. This is not an activity that an employee should be assigned to lightly but should be evaluated thoroughly to identify employees who are most capable of executing this responsibility.

Management should effectively prepare and support employees in this effort. Employees should be properly prepared in advance of the audit through training and program implementation activity. Employees should be coupled with programs they come into contact within their daily job. For example, a maintenance technician who routinely repairs storage racking on the facility production floor through the use of welding could be given responsibility for the Hot Work Program. The employee may naturally know a great deal of the information related to the program, such as:

- Where hot work permits are needed
- The process used to issue hot work permits
- Who can issue hot work permits
- Safety precautions that must be taken

This information may have been gained through training that the employee has received. However, management may need to supplement this knowledge with additional information that addresses how the program is administered. This information might include:

- The qualifications for conducting hot work training
- Where the two-part hot work permits are procured
- Vendors utilized to repair and maintain the equipment used within the Hot Work Program

With the complete spectrum of information at hand, the employee can then adequately respond to program questions asked by the auditor. The employee can also direct the auditor to written program content and documentation that supports the implementation of the Hot Work Program.

FACILITY INSPECTION

The facility inspection provides another way in which employees can become engaged in the audit process. Rather than a member of management escorting the auditor through the facility, employees in each of the departments can be assigned to guide the auditor through their work areas. The employees working in the departments will have an intimate working knowledge of the processes and procedures that are used in the department and can be a great source of information for the auditor.

Senior employees who have demonstrated responsibility in their position can engage by becoming the primary point of contact point for the facility inspection in their department. Like employees who wish to engage in the documentation review portion of the audit, employees who wish to engage in the facility inspection portion should be thoroughly prepared. They should be equipped with appropriate information, such as:

- The nature of the inspection
- The scope of the inspection
- Questions the auditor might ask during the inspection
- Freedom to interact with the auditor without fear of repercussion

Employees can then effectively act as a guide for the facility inspection of their respective departments. Personal knowledge of the work performed in the department combined with preparation of the audit process should make them feel comfortable interacting with the auditor. If employees feel like they have been asked a question to which they cannot respond effectively, they should be made to feel like they can contact the Department Supervisor for assistance.

The engagement of employees in the process has several benefits. First, it places the employees in a position of being part of the process. Rather than being spectators, they have an active role in the audit through making recommendations, implementing solutions, assisting with the documentation review, or by accompanying the auditor in the facility during the inspection. Second, it can be impressive to the auditor that the programs within the scope of the audit have been so internalized by the facility that employees are able to accept responsibilities in the execution of the audit. Though managers may typically be the primary points of contact for auditors, it can be a refreshing experience for the auditor to engage with employees.

CASE STUDIES

FIRE

Wes is responsible for managing the upcoming annual audit at his station. He has reviewed the process with each shift and explained to all firefighters and officers what to expect. Allen is a new recruit and is anxious to get involved in everything possible. He approaches Allen to volunteer his services in getting prepared for the audit. He acknowledges that he is new to the department but would like to become involved.

SAFETY

Dean has been the facility Safety Manager for five years and he has worked to manage the audit process each year. Rather than continually depending on management to facilitate the audit, he would like to involve employees. Though employees have never been asked to be directly involved with the audit, Dean believes that Mike, a Maintenance Technician, would be an asset to the process. Mike has worked for the company for 15 years and has worked in every aspect of facilities maintenance. Dean decides to approach Mike with the idea of leading aspects of the audit that affect maintenance.

SECURITY

As the site Security Manager, Tina has been working on correcting deficiencies from the previous year's audit. This year's audit is two months away and she needs to find solutions for some of the findings. She decides to approach the Safety Committee for assistance because each of her issues involves exposure to workplace violence. She contacts the Chairperson of the Safety Committee who agrees to have her attend next week's meeting. In a brief conversation with the Chairperson, she finds that the Safety Committee is comprised of ten employees from each department who range in tenure with the company from one year to ten years. In addition to the employees, there are two department supervisors on the committee.

- How might the employees in these situations become involved in the audit that will be conducted?
- What influence does their level of experience have on the degree to which they can become involved in the audit and addressing audit findings?

EXERCISES

For the following questions, identify a single facility environment in which you would like to situate your responses and answer each question accordingly. Answer each question in the context of a fire, safety, or security audit.

1. In what ways does employee involvement add value to the audit process?
2. In what way might employees make recommendations to assist in achieving a successful outcome of an audit?
3. How can a Safety Committee assist in achieving a successful outcome of the audit?
4. What type of employee might you look for to become engaged in the documentation review portion of an audit?
5. What type of employee might you look for to become engaged in the facility inspection portion of an audit?

29 Challenges

Creating and implementing an audit program is a difficult task. Though the auditing process can have a great deal of value to an organization, there are a few challenges that might be encountered. An understanding of the nature of these challenges can help to identify a strategy to address them if they surface throughout the audit program development and implementation process. Challenges that might be encountered include:

- Commitment from key stakeholders
- Budget
- Ill-prepared personnel
- Technical problems
- Lack of knowledge regarding certain issues

COMMITMENT

Commitment from key stakeholders will be necessary for an audit program to be successful. Key stakeholders include:

- Chief Executive Officer (CEO)
- Divisional Vice Presidents
- Regional Managers
- Plant Managers
- Shift Managers
- Maintenance Managers
- Fire Chiefs
- Department Officers
- Training Officers

Each of these individuals will be significantly engaged in the audit or will have an interest in the outcome of the audit. Their commitment and ongoing support will be critical to the success of the program. It will be necessary to gain an understanding of how the audit program will meet the needs of these individuals.

One example might be in financial loss that is occurring in the organization. This loss could include:

- Worker's compensation cost due to injuries
- Lower quality due to replacing a skilled employee who is away from work due to an injury with a worker who is less skilled
- Dual personnel cost through paying an injured employee who is home recovering from an injury on lost time and paying for a second employee or temporary worker to continue production in the injured employee's absence

DOI: 10.1201/9781003371465-36

- Repair and replacement of equipment due to property damage incidents
- Down time that results from damaged equipment
- Theft of product
- Embezzlement of company funds

The goals of organizational management are to maintain operational capacity, deliver on production goals, and provide a return on shareholder investment. All the losses mentioned in the list above hamper management's ability to achieve these goals. An audit program is a tool that can assist organizational leaders by thoroughly examining the system that is in place at a given facility and identify where opportunities for improvement exist as well as where best practices are occurring. Identifying opportunities for improvement can have a positive impact by lowering exposure to loss. Sharing best practices is an opportunity to strengthen the integrity of the system across the organization by implementing such practices in all facilities. Helping stakeholders understand these losses, how they prohibit the achievement of organizational goals, and the value of the audit program in preventing these losses can facilitate the development and implementation of the audit program. This can result in organizational leaders committing to and supporting the audit program.

BUDGET

The implementation of an audit program can require significant resources in the organizational budget. Like many other budget categories within an organization, there are several things that will need to be included in the budget to ensure the successful implementation of an audit program. Such budget items might include:

- Contracted services or internal time dedicated to the development of the audit program and audit document
- Hiring or assigning auditors
- Time required to train auditors
- Travel expenses
- Purchase of company vehicles for regional travel
- Equipment, such as laptops and cell phones
- Allocating and equipping office space for auditors
- IT support cost for the system identified to create the audit

If the first challenge of commitment to the audit program is not achieved, it will be more difficult to achieve the goal of allocating budget to the audit process. By first gaining the commitment of organizational leadership to the audit program, it will be somewhat easier to obtain the budget that is requested. However, there will still be the challenge of getting everything that is requested. Money-conscious organizations will scrutinize each dollar that is requested and spent, so detailed planning and justification for each budget item will be necessary to obtain the budget that is requested. It will be important to connect budget items to a return on investment wherever possible. This return on investment could be direct or indirect savings to the organization.

Once a budget is established, a subsequent challenge is to manage the budget responsibly. Ethical conduct in budget management has been highlighted through the Sarbanes-Oxley Act which was created to legally mandate financial stewardship in publicly held organizations. Money must be spent and accounted for in keeping with organizational finance and accounting policies. For example, a travel policy might be in place that dictates certain thresholds of expenditures for food, lodging, and travel. Auditors will need to comply with such a policy by spending travel money responsibly. The audit program manager will need to assume control of the budget and closely manage expenditures to ensure there is no appearance of impropriety.

ILL-PREPARED PERSONNEL

The responsibility of an auditor in preparing a location for an audit might include such activities as:

- Notifying the facility of the date of the audit
- Making the facility aware of what to expect in the audit process
- E-mailing a form that includes an audit timetable and documentation that will be asked for during the audit
- Following up periodically with the location to ensure questions are answered prior to arriving to conduct the audit

A great deal of effort may go into preparing the facility for the upcoming audit. Yet when the auditor arrives, she finds that:

- Her point of contact cannot meet with her due to needing to respond to a long list of e-mails
- The requested management system documentation has not yet been collected
- A work area has not yet been designated for her to set up her computer and begin work on the audit

The facility is ill-prepared for the audit. At this point the auditor will need to make a choice. She will need to forge ahead and conduct the audit as scheduled or she will need to terminate the audit due to the lack of preparedness of facility personnel and reschedule the audit for a time when the facility is better prepared.

The benefit of moving forward with the audit is that travel money that has been spent will not be wasted due to the audit being conducted as planned. One perspective of moving forward with the audit is that the facility's performance will simply be assessed as it exists, regardless of how well facility personnel are prepared. The audit acts as an assessment in time of how the facility is performing and could be to their failure that they did not effectively prepare for the audit.

The benefit of canceling the audit and the auditor returning home is that the effort to conduct the audit will not be wasted. The audit requires a certain amount of logistical support and interaction from facility management for the audit to be a meaningful experience. If it is obvious that the support and interaction will not be provided, the decision might be to cancel the audit and reschedule it for the near future when the facility can dedicate the time and resources necessary to conduct the audit.

The challenge of an ill-prepared facility has no simple answer. The auditor may need to make a judgment based on the evidence that is witnessed upon arrival. Though it may be difficult to do, canceling the audit due to facility management being ill-prepared might be the best course of action. This action might result in negative communication from upper management to the facility for lack of preparedness, but a positive result can be the message that such behavior will not be tolerated within the scope of the audit program.

TECHNICAL PROBLEMS

A laptop and connection to the Internet are fundamental tools necessary to conduct audits. A laptop or tablet is needed to access the audit document that has been created, whether it be a Microsoft Excel document or audit product provided through an online platform. The audit document will be populated with scoring assessments, comments, and digital photos. The Internet and integrity of the audit document product are necessary to conduct the audit. Both tools yield the possibility for technical problems to arise.

IT support is a critical component in responding to technical issues. The auditor will need to be aware of whom to contact if software, hardware, or Internet connection problems surface during an audit. Contact numbers for IT support personnel will need to be maintained to expedite the resolution of problems that might be experienced while onsite at the facility. If the problem cannot be resolved in a timely manner, a hard copy backup of the audit document will be helpful in allowing the audit to continue. Though the auditor will have to do more work to complete the audit report upon returning to the office, the hard copy will allow the auditor to complete the audit in the event IT support cannot resolve problems that have surface.

LACK OF KNOWLEDGE

A fire, safety, or security audit will cover a wide spectrum of information. A great deal of work will be done to ensure the audit addresses all the variables that exist at each of the facilities being audited. A challenge that an auditor might encounter is a unique application or environmental issue in which they are not sure how a given component of the audit should apply. For example, an auditor might be in the process of conducting a facility inspection when he notices a confined space. The confined space has unique attributes with which the auditor is not knowledgeable. It is acceptable for the auditor to simply state, "I am not sure." rather than make an immediate judgment that is in error. The auditor can utilize the time that is dedicated to creating the audit report to research the issue and make an educated judgment. The auditor can then contact facility management to inform them of the outcome of the issue.

It is not acceptable to expect facility management to research the issue and count it against the facility in the scoring of the audit. The auditor is expected to be the subject matter expert on issues that are within the scope of the audit. If the auditor is not familiar with the application of a certain company policy or regulation to something that is encountered during the audit, there can be no expectation that facility personnel will be able to effectively research and identify a proper course of action. The auditor will need to take the initiative to research the issue and report the findings to location management.

CASE STUDIES

FIRE

Tim is conducting the annual audit at Station 2 and is accompanied by Sue, a newly promoted Captain, on the facility inspection. Tim notices a means of egress issue and calls it to Sue's attention. Having helped in the preparation for the audit, Sue is surprised at the finding. After thinking about the issue, she contends that Tim is thinking of a requirement that is under a previous edition of NFPA's Life Safety Code.

SAFETY

Cal is accompanying the auditor, Shirley, on the facility inspection. They come upon a confined space that Shirley recognized as having been recorded in the facility Confined Space Entry Program. She made a note to examine the space because the facility records indicated that the confined space had been entered, but that an entry permit had not been utilized. She asks Cal why a permit was not utilized. Cal responds with an explanation that included a review of work that was conducted and that there were no hazards in the environment that would cause it to require the use of an entry permit and the subsequent Emergency Response Team. Shirley does not agree and states that a permit should have been used but is vague on the exact hazard that would have been present during the time of the entry.

SECURITY

Veronica arrives at the facility to conduct the annual security audit. She has worked with the facility over the course of the previous months to prepare them for the event. Upon her arrival, she is asked to wait in the front lobby until Bob, her point of contact, can meet with her. One hour later Bob walks into the lobby and apologizes for the delay. He explains that he must now go into a meeting that will last the remainder of the afternoon. He also apologizes by stating that they have been unable to collect the written program material that was requested in previous e-mails and telephone calls to be ready upon her arrival. The audit was scheduled to take place between this afternoon and all the next day. She realizes that a great deal of this time is in jeopardy of being lost.

- What is the nature of the challenge that exists in these scenarios?
- How should the auditor respond?

EXERCISES

For the following questions, identify a single facility environment in which you would like to situate your responses and answer each question accordingly. Answer each question in the context of a fire, safety, or security audit.

1. What resources are available to gain organizational leadership commitment for an audit program?
2. What items might you include in a budget for an audit program? How would you go about obtaining the money that you request for each item in your budget?
3. What would be your strategy to address the potential of technical problems occurring during an audit?
4. Is it acceptable for an auditor to admit a lack of knowledge regarding a certain issue during an audit? Why or why not?
5. What is the greatest challenge that you see as being an issue in carrying out your audit program?

30 Subsidiary Organization Implications

The functionality of a subsidiary relationship with a parent company can vary greatly. Relationships can range from the parent company having little involvement in the operation of the subsidiary to the parent company having a great deal of involvement in the operation of the subsidiary. This creates a unique lens through which each subsidiary relationship should be viewed. The degree of relationship could impact the degree of risk assumed by the parent company. For example, a parent company may choose to be highly involved in the auditing and assessment of the subsidiary organization, which could transfer a great deal of risk to the parent company by having a high level of direction of the subsidiary organization and its daily activities. Subject matter experts employed by the parent company might be sent into the subsidiary organization to conduct robust management system audits and might serve as a resource to provide guidance on performance improvement. Such a decision might be made where the subsidiary organization does not have the resources needed to meet the performance standards of the parent organization. Conversely, the parent organization might choose to engage in a relationship where the subsidiary organization functions autonomously and is expected to conduct internal assessments utilizing resources within or contracted by the subsidiary organization to meet the performance expectations of the parent company. These two models and potential variations should be explored with legal counsel to understand and determine the level of risk the parent company will choose to accept based on unique variables involved.

A due diligence inspection is a form of initial audit a parent company will engage in when considering the final purchase of a subsidiary organization. Internally employed or contracted subject matter experts will be sent to the facility to conduct an analysis of the current state of operations of the facility and where gaps might exist based on performance standards held by the parent company. For example, the safety component of a due diligence inspection might reveal the need to invest significant amounts of money to improve the physical environment of the facility, such as the need to improve walking and working surfaces and machine guarding. Improvements can be made by the parent company and subsidiary organization partnering on how to move forward with needed improvements. Responsibilities of the parent company and subsidiary organization will be defined.

Once the procurement of a subsidiary organization occurs, the decision might be to allow the subsidiary organization to continue operating independently. The subsidiary organization might maintain its own central office, operating budget, management structure, and retain its organizational name. Such independence might be evidenced in several ways:

DOI: 10.1201/9781003371465-37

- Subsidiary organization leadership might insert authority and manage compliance inspections
- Subsidiary organization might engage its own in-house counsel versus accessing that of the parent company
- Parent company having limited access to an incident scene due to being fully managed by subsidiary organization subject matter experts
- OSHA might recognize and only engage individuals of the subsidiary organization in a closing conference
- In ongoing compliance events, OSHA might recognize and cite the subsidiary organization for violations as opposed to the parent company
- Subsidiary organization management structure that is capable of fully functioning independently of the parent company
- Subsidiary organizations operating its own workers' compensation management and payment system

Closer partnerships might be formed between the parent company and the subsidiary organization. The parent company might choose to provide more direct support to the subsidiary company. This might be evidenced through parent company activities such as:

- Conducting risk assessments to determine what measures are needed to permit safe operation
- Conducting management-led safety training to inform subsidiary workers and managers of management system activities that must be performed
- If worker safety could not be achieved through alternative measures, the parent company might suspend operation until workers could be safely introduced into its operation
- Provide subject matter expertise support and direction to the subsidiary organization
- Provide funding to address management system improvements within the subsidiary organization

The relationship established between the parent company and the subsidiary organization will have an impact on how ongoing management system assessment is conducted and communicated. Effort will need to be made to clearly define who will engage in the following activities within the parent company and/or subsidiary organization:

- Determine the scope of management system activity
- Assign responsibilities within the management system
- Define tools that will be used to assess the management system
- Determine who will conduct each assessment
- Determine the frequency at which each assessment tool will be utilized
- Determine how results of the assessments will be communicated to personnel within the subsidiary organization and parent company
- Establish levels of priority to which each assessment finding will be assigned

- Establish the process to be used to address assessment findings
- Determine how each finding is brought to closure

Acquisitions will require forethought as to how the partnership will exist and how ongoing activities related to management systems will be carried out and assessed. The level of risk and integration into corporate operations can have an impact on the degree to which the parent company chooses to engage in and provide support to the subsidiary organization. An effective strategy can be developed to ensure the subsidiary organization develops and performs to the level mandated by the parent company, whether it involves direct interaction with the parent company, or the subsidiary organization is to function autonomously.

CASE STUDIES

FIRE

Sue is the director of a private organization that provides supplemental fire protection services to individuals and communities. A recent rise in wildland fires has caused her organization to consider expanding its business. She has identified a company that is struggling financially to maintain its operations but might be a good fit for the culture and direction of her organization.

SAFETY

Mark is the safety director of the US division of an international agricultural company. The US division is considering the purchase of a smaller company to expand their presence and operation in the southeastern US. The organization under consideration appears to be profitable, but a review of OSHA reports indicates a history of violations in multiple states in which the organization has facilities.

SECURITY

Duke has been the Corporate Security Manager for his company that operates facilities in the northeastern part of the US. His company would like to expand its operation along the eastern coast and has found a potential company to purchase. The acquisition of the company would allow them to provide operations in areas that are desirable for the development of their business model. The challenge is that the acquisition would place them in primarily urban areas compared to most of their current locations being in suburban and rural areas.

- What steps should be taken to manage the risk of purchasing the subsidiary organization?
- What issues should be considered in determining if the parent company should become highly involved in the subsidiary organization or if the subsidiary organization should be allowed to operate autonomously?

EXERCISES

For the following questions, identify a single facility environment in which you would like to situate your responses and answer each question accordingly. Answer each question in the context of a fire, safety, or security audit.

1. What risks might be present if a parent company chooses to exercise substantial control over a subsidiary organization?
2. Why might a parent company choose to allow a subsidiary organization to operate autonomously?
3. What role does a due diligence inspection play in the assessment of purchasing a subsidiary organization?
4. What evidence might support the segregation in operations between a parent company and a subsidiary organization?
5. What evidence might support a close operational relationship between a parent company and a subsidiary organization?
6. What elements will need to be defined in determining the relationship between a parent company and subsidiary organization?

Appendix A: OSHA Policy on Self-Audits

Reference: https://www.osha.gov/laws-regs/federalregister/2000-07-28

DEPARTMENT OF LABOR

OCCUPATIONAL SAFETY AND HEALTH ADMINISTRATION

[Docket No. W-100]

FINAL POLICY CONCERNING THE OCCUPATIONAL SAFETY AND HEALTH ADMINISTRATION'S TREATMENT OF VOLUNTARY EMPLOYER SAFETY AND HEALTH SELF-AUDITS

AGENCY   Occupational Safety and Health Administration, USDOL.

ACTION   Notice of final policy

Authority   Sec. 8(a) and 8(b), Pub. L. 91-596, 84 Stat. 1599 (29 U.S.C. 657)

SUMMARY   The Occupational Safety and Health Administration (OSHA) has developed a final policy describing the Agency's treatment of voluntary employer self-audits that assess workplace safety and health conditions, including compliance with the Occupational Safety and Health Act (Act). The policy provides that the Agency will not routinely request self-audit reports at the initiation of an inspection, and the Agency will not use self-audit reports as a means of identifying hazards upon which to focus during an inspection. In addition, where a voluntary self-audit identifies a hazardous condition, and the employer has corrected the violative condition prior to the initiation of an inspection (or a related accident, illness, or injury that triggers the OSHA inspection) and has taken appropriate steps to prevent the recurrence of the condition, the Agency will refrain from issuing a citation, even if the violative condition existed within the six month limitations period during which OSHA is authorized to issue citations. Where a voluntary self-audit identifies a hazardous condition, and the employer promptly undertakes appropriate measures to correct the violative condition and to provide interim employee protection, but has not completely corrected the violative condition when an OSHA inspection occurs, the Agency will treat the audit report as evidence of good faith, and not as evidence of a willful violation of the Act.

FOR FURTHER INFORMATION CONTACT   Richard E. Fairfax, Occupational Safety and Health Administration, Directorate of Compliance Programs, Room N-3603, U.S. Department of Labor, 200 Constitution Avenue, NW, Washington, DC 20210, Telephone: 202-693-2100.

SUPPLEMENTARY INFORMATION

I.   BACKGROUND INFORMATION

On October 6, 1999, OSHA published a "Proposed Policy Statement Concerning the Occupational Safety and Health Administration's Use of Voluntary Employer Safety and Health Self-Audits" in the **Federal Register**. 64 FR 54358 (1999). The policy statement described the Agency's proposal regarding the manner in which it would treat voluntary employer self-audits that assess workplace safety and health conditions, including compliance with the Act. The proposed policy statement provided that the Agency would not routinely request voluntary employer self-audit reports at the initiation of an inspection. Further, the proposed policy provided that, where an employer identified a hazardous condition through a voluntary self-audit, and the employer promptly undertook appropriate corrective measures, OSHA would treat the audit report as evidence of good faith, and not as evidence of a willful violation. It was, and remains, the Agency's intention to develop and implement a policy that recognizes the value of voluntary self-audit programs that are designed to allow employers, or their agents, to identify and promptly correct hazardous conditions. In limited situations, however, documentation related to voluntary self-audits plays an important role in the Agency's ability to effectively and faithfully carry out its inspection and enforcement obligations under the Act.

Although the Agency is not required by the Administrative Procedures Act, 5 U.S.C. 551, **et seq.,** to engage in notice and comment rulemaking procedures prior to the adoption and implementation of this policy, OSHA requested public comment regarding its proposed policy statement in order to gain input and insight from employers, employees, employee representatives, and other interested parties. OSHA received and thoroughly reviewed comments from a variety of sources. The Agency has modified the proposed policy to incorporate those comments that further OSHA's dual purposes in proposing the voluntary self-audit policy – **i.e.,** to provide appropriate, positive treatment that is in accord with the value that voluntary self-audits have for employers' safety and health compliance efforts, while maintaining the Agency's authority to gain access to voluntary self-audit documentation in limited circumstances in which access is important to effectively and faithfully enforce the Act. The Agency has not incorporated those comments that it considered to be contrary to its purposes in proposing this policy or that it considered to be beyond the scope of its intent in proposing the policy.

II.   SUBSTANTIVE MODIFICATIONS TO THE PROPOSED POLICY

Based upon input that the Agency received from interested parties, OSHA has made several substantive changes to its proposed policy on the treatment of voluntary employer self-audits.

1.   Modifications to Certain Definitions in the Policy

In the final policy, the Agency has defined the term "self-audit" to include health and safety audits conducted for an employer by a third party. In addition, in defining the terms "systematic" and "documented," the Agency has added the words "or for" before the phrase "the employer" to clarify that an audit conducted by a third party for an employer is covered by the final policy. OSHA values the role that independent safety and health professionals play in furthering occupational safety and health and encourages employers to utilize their services when appropriate.

The Agency has changed the definition of the word "objective" by deleting reference to "safety and health professional[s]" and by broadening the class of persons who may conduct an "objective" self-audit to include competent employees and management officials. Thus, in the final policy, a self-audit is "objective" if it is conducted "by or under the direction of an individual or group of individuals who are competent to identify workplace safety or health hazards, given the scope and complexity of the processes under review." This modification is responsive to suggestions from small business employers, organizations such as the National Advisory Committee on Occupational Safety and Health, and other members of the public. Employers, particularly small business employers, who might not have the financial resources to hire an independent consultant, may use their own personnel who do not have professional certification, but who do have the necessary experience or training to conduct an effective and thorough self-audit. In addition, the Agency recognizes the expertise that many joint labor-management safety committees have developed with respect to workplace safety and health issues and acknowledges that audits conducted by such committees should qualify for recognition under this policy.

2. Training for Compliance Safety and Health Officers

In the final policy, the Agency has added the following statement: "All OSHA personnel applying this policy will receive instruction in order to ensure the consistent and appropriate application of the policy." The Agency received comments from employers expressing their concerns regarding the potential for inconsistent implementation and application of the policy. OSHA agrees that an effective policy can be achieved only through consistent implementation and application. Thus, in the final policy, OSHA has explicitly stated that training will be provided, over a period of time, to all personnel who will apply this policy in order to ensure its consistent and proper implementation.

3. Citation Policy for Violative Conditions Identified and Corrected Through Voluntary Self-Audits

In response to numerous suggestions from commenters, the Agency has added a provision explicitly stating that OSHA will not issue citations for violative conditions discovered during a voluntary self-audit and corrected prior to the initiation of an inspection (or a related accident, illness, or injury that triggers the inspection), even if the violative condition existed within the six month limitations period during which OSHA is authorized to issue citations. OSHA encourages employers to conduct voluntary self-audits and to promptly correct all violations of the Act that are discovered in order to ensure safety and health in the workplace. Thus, in the final policy, the Agency has incorporated its current enforcement practice and will refrain from issuing a citation for a violative condition that an employer discovered as a result of a voluntary self-audit, if the employer corrects the condition prior to the initiation of an OSHA inspection (or a related accident, illness, or injury that triggers the OSHA inspection), and if the employer has taken appropriate steps to prevent a recurrence of the violative condition.

4. Employers' Prerogative to Voluntarily Provide Self-Audit Documentation

Several parties requested that the Agency provide that, in those situations in which the Agency has not requested or used voluntary self-audit documentation in conducting its inspection, employers be permitted to take advantage of the policy by providing the Agency with evidence of their voluntary self-audit program. Since OSHA inspectors rarely request voluntary self-audit documentation when conducting inspections, and this policy states the Agency's intent that inspectors should request such documentation only in limited situations, OSHA recognizes that there will be a significant number of instances in which the Agency is unaware of an employer's voluntary self-audit activities, and thus the employer would not be considered for recognition under the policy. Therefore, the final policy provides that an employer voluntarily may provide the Agency with self-audit documentation, and the employer may be eligible to receive the benefits that are detailed in this policy.

III. COMMENTS NOT INCORPORATED INTO FINAL POLICY

A number of parties offered comments that have not been included in the final policy. While the Agency considered thoroughly each of the comments that it received, OSHA considered the following comments either to be inconsistent with the Agency's dual purposes in proposing the policy or to be beyond the scope of the proposed policy.

1. Employee Participation in the Voluntary Self-Audit Process

Two union representatives maintained that OSHA should require employers to disclose self-audit results to their employees and their representatives and that OSHA should not grant good faith credit to any employer who has not disclosed all of the audit results both to OSHA and to its employees. OSHA agrees that the interests of workplace safety and health are advanced when employers share self-audit results with employees and employee representatives. However, because this is not a rulemaking procedure, the Agency considers it to be inappropriate to use this policy to adopt a practice that may be deemed to modify the legal duties of employers. Moreover, insofar as the purpose of this statement is to clarify current OSHA practices and to provide appropriate, positive treatment that is in accord with the value of voluntary self-audits, the Agency believes that it may be counterproductive to impose additional requirements on employers in order to qualify for inclusion under the policy.

2. More Significant Proposed Penalty Reductions

Several parties suggested that OSHA should provide a more significant proposed penalty reduction for an employer's "good faith" by offering proposed penalty reductions in excess of 25 percent to employers who identify violative conditions during voluntary self-audits and who have begun to correct the conditions, but who have not completed abatement prior to the initiation of an OSHA inspection. OSHA's current guidelines account for an employer's "good faith" when the Agency calculates a proposed penalty for a violation of the Act. These guidelines allow a penalty reduction of up to 25 percent in recognition of an employer's "good faith," if the employer has developed and implemented a written health and safety program, which provides for appropriate management commitment and employee involvement; worksite analysis for the purpose of hazard identification; hazard prevention and control measures; and safety and health training. The Agency has stated that it will treat a voluntary self-audit, which results in prompt corrective action and appropriate steps to prevent similar violations, as strong evidence of the employer's good faith with respect to the matters covered by the voluntary self-audit. However, a voluntary self-audit is only one of the many steps that employers can and do undertake to protect the health and safety of their employees, and OSHA does not believe that the goals of the Act would be furthered by an additional "good faith" penalty reduction that is keyed directly and exclusively to voluntary self-audits. Rather, the Agency believes that its current "good faith" penalty reduction provisions, in conjunction with the inherent advantages that employers gain by conducting voluntary self-audits and the treatment that this policy provides for voluntary self-audits, provide appropriate, positive recognition for voluntary self-audits.

3. Total Prohibition Against the Use of Voluntary Self-Audit Documentation

Many employers and employer associations stated that OSHA should refrain totally from using voluntary self-audit information as a part of the Agency's enforcement efforts under the Act. The Agency has not incorporated this comment into its policy because it believes that a complete prohibition is unnecessary in order to provide appropriate, positive treatment for voluntary self-audits. In addition, the Agency believes that, in some circumstances, a complete prohibition would prevent it from effectively enforcing the Act.

The implementation of this policy will publicly state the Agency's policy to request voluntary self-audit documentation only in limited situations. A substantial number of employers already conduct voluntary self-audits for their own benefit and for the benefit of their employees. The Agency believes that this policy, with its explicit provisions concerning the Agency's use of voluntary self-audit documentation, will provide the assurances that additional employers may need in order to conduct voluntary self-audits. Indeed, under the policy, employers who respond promptly and appropriately to hazardous conditions that are identified in a voluntary self-audit can only be rewarded for having conducted the self-audit.

On the other hand, there are legitimate circumstances in which voluntary self-audit data are important to enable the Agency to effectively enforce the Act. For example, such information may allow an inspector, who has already identified a hazard, to determine the scope of the hazard or to assess the manner in which the condition can be abated. In addition, pursuant to Occupational Safety and Health Review Commission precedent, the Secretary of Labor has the obligation to demonstrate that an employer had knowledge of a cited violative condition, and, in certain situations, the obligation to demonstrate that an employer was so indifferent to recognized occupational health or safety hazards that more significant penalties are justified in order to effectuate the provisions of the Act. Thus, the Agency believes that a complete prohibition against the use of voluntary self-audit documentation would be an imprudent policy because it would hamper OSHA's ability to enforce the Act effectively.

4. More Precisely Defined Limitations on the Agency's Use of Voluntary Self-Audit Documentation

In the proposed policy statement, the Agency had proposed to "refrain from routinely requesting reports of voluntary self-audits at the initiation of an enforcement inspection." OSHA explained that it intended to seek access to such reports only in limited situations in which the Agency had an independent basis to believe that a specific safety or health hazard warrants investigation, and had determined that such records may be relevant to identify or determine the circumstances or nature of the hazardous condition. However, several employers asked that the Agency more precisely detail the specific situations in which its inspectors may request voluntary self-audit documentation.

The Agency has decided not to attempt to modify its proposed policy in this manner for several reasons. First, OSHA believes that, given the diversity of circumstances that inspectors encounter in conducting thousands of workplace inspections each year, it is not feasible to comprehensively list or to describe with any specificity each of those situations in which it would be appropriate for an inspector to request voluntary self-audit documentation. Rather, the Agency believes that the implementation of this policy will provide sufficient specificity to assure employers that inspectors will seek voluntary self-audit documentation only in limited and generally defined situations. Second, OSHA recognizes the skill and experience of its inspectors and believes that it is essential for the Agency and its inspectors to have some discretion in implementing this policy in order to effectively and efficiently fulfill the Act's mandate to detect and identify occupational safety and health hazards. Third, in refraining from an attempt to more specifically define those discrete circumstances in which inspectors may request voluntary self-audit documentation, the Agency has adopted the comment offered by several employers and their representatives who expressed concern that such specificity may in practice increase the frequency with which inspectors request voluntary self-audit documentation, given the natural human inclination to interpret specific examples as situations in which a request for self-audit documentation is mandated, as opposed to merely permitted, pursuant to the policy.

5. Adoption of a Formal Rule Regarding the Agency's Treatment of Voluntary Self-Audits

Several commenters suggested that the Agency should adopt the "Final Policy Concerning the Occupational Safety and Health Administration's Treatment of Voluntary Employer Safety and Health Self-Audits" as a formal rule that would be legally binding on the Agency. However, OSHA has declined to incorporate this comment and believes that the policy, as adopted, provides sufficient assurance that employers who conduct voluntary self-audits, and who take prompt and appropriate steps to address occupational hazards that are identified in such audits, will not be penalized by OSHA for conducting voluntary self-audits. In addition, since this policy is an internal policy that is intended only to provide OSHA inspectors with guidance regarding the circumstances under which the Agency considers it appropriate to review and consider documentation generated by employers as a result of voluntary self-audits, the Agency believes it is imprudent and unnecessary to expend the time, money, and other resources required to promulgate a formal rule. Finally, the Agency believes that a rule that creates legal rights for third parties would be more likely to produce unproductive litigation than will a policy that only provides guidance to OSHA inspectors. This type of litigation would not further the health and safety purposes of the Act.

6. No Citation for Partial or Planned Correction of Violative Conditions Identified through a Voluntary Self-Audit

A number of employers stated that the Agency should refrain from issuing a citation in any situation in which an employer has identified a hazardous condition and is in the process of correcting that condition, or has developed a plan or program for correcting that condition, at the time that OSHA conducts an inspection of the employer's facility. OSHA has decided not to incorporate this comment into the final policy for several reasons. First, the agency recognizes that the prompt correction of hazardous workplace conditions is essential for the prevention of occupational illnesses, injuries, and fatalities. The Agency is concerned that a policy that excuses an employer for an abatement plan alone, or for abatement actions that do not constitute the complete elimination of the hazard, may serve to diminish an employer's incentive to promptly and completely eliminate workplace hazards. Second, the Agency believes that such a policy would be inconsistent with the Act's mandate, which is to assure, so far as possible, safe and healthful working conditions for every working man and woman in the Nation. In enforcing the Act, OSHA only issues citations in cases in which employees actually are exposed to hazards associated with violative conditions. While the final policy recognizes that employers who identify hazardous conditions through the use of voluntary self-audits, and are in the process of correcting those hazards, may deserve a "good faith" reduction in the penalty that OSHA proposes for the violation, the Agency does not believe that the Act contemplates that OSHA will refrain totally from issuing citations in situations in which employees are working in an environment in which they are exposed to serious occupational hazards.

IV. DESCRIPTION OF THE FINAL POLICY

The policy applies to audits (1) that are systematic, documented, and objective reviews conducted by, or for, employers to review their operations and practices to ascertain compliance with the Act, and (2) that are not mandated by the Act, rules or orders issued pursuant to the Act, or settlement agreements. A systematic audit is planned, and it is designed to be appropriate to the scope of the hazards that it addresses and to provide a basis for corrective action. Ad hoc observations and other ad hoc communications concerning a hazardous condition made during the ordinary course of business are not included within the definition of a "self-audit" or "voluntary self-audit report." The findings resulting from the systematic self-audit must be documented contemporaneously (at the time the condition is discovered or immediately after completion of the audit) so as to assure that they receive prompt attention.

The self-audit also must be conducted by or supervised by a competent person who is capable of identifying the relevant workplace hazards. Employees or management officials who have the training or experience that is necessary to identify workplace safety or health hazards, given the scope and complexity of the processes under review, are considered to be competent persons under the policy, even though they may not maintain engineering, scientific, industrial hygiene, or other relevant professional accreditation.

In order to qualify for inclusion under the policy, a self-audit need not review or analyze an entire plant, facility, or operation. For example, a voluntary self-audit designed to identify hazards associated with a particular process or hazard (as opposed to an entire plant, facility, or operation) will qualify for consideration under the policy.

The policy provides that OSHA will not routinely request voluntary self-audit reports when initiating an inspection, and that the Agency will not use voluntary self-audit reports as a means of identifying hazards upon which to focus during an inspection. Rather, OSHA intends to seek access to such reports only in limited situations in which the Agency has an independent basis to believe that a specific safety or health hazard warrants investigation, and has determined that such records may be relevant to identify or determine the circumstances of the hazardous condition. For example, an inspector might seek access to self-audit documentation following a fatal or catastrophic accident when OSHA is investigating the circumstances of the accident to assess compliance and to assure that hazardous conditions are abated. Likewise, it would be consistent with this policy to request self-audit documentation when the Agency has an independent basis for believing that a hazard exists. The Agency believes that this provision is responsive to the concerns of employers who sought assurances that OSHA would not use voluntary self-audit documentation during an inspection as a "road map" to identify violations of the Act.

OSHA emphasizes that it is not seeking through this policy to expand the situations in which it requests production of voluntary self-audit reports beyond its present practice. In addition, OSHA intends to seek access only to those audit reports, or portions of those reports, that are relevant to the particular matters that it is investigating.

OSHA has defined "voluntary self-audit report" to include information obtained in the audit, as well as analyses and recommendations. The effect is to include audit information in the documents that OSHA will not routinely request at the initiation of the inspection. OSHA has defined the term this way because the Agency believes that the definition responds to the concerns raised by employers about the effect of routine OSHA requests for voluntary self-audit findings.

The policy also contains provisions designed to assure that employers who respond with prompt corrective actions will receive corresponding benefits following an OSHA inspection. These provisions would come into play when OSHA obtains a voluntary self-audit report, either because the employer has voluntarily provided it to OSHA, as commonly occurs, or because OSHA has required production of the report. In response to public comment, OSHA has expressly stated in the final policy that employers may voluntarily provide OSHA with self-audit documentation and that those employers may be eligible to receive the benefits detailed in the policy.

The policy explains that OSHA will refrain from issuing a citation for a violative condition that an employer has discovered through a voluntary self-audit and has corrected prior to the initiation of an inspection (or a related accident, illness, or injury that triggers the inspection), if the employer also has taken appropriate steps to prevent the recurrence of the condition. In situations in which the corrective steps have not been completed at the time of the inspection, OSHA will treat the voluntary self-audit report as evidence of good faith, not as evidence of a willful violation, provided that the employer has responded promptly with appropriate corrective action to the violative conditions identified in the audit. Accordingly, if the employer is responding in good faith and in a timely manner to correct a violative condition discovered in a voluntary self-audit, and OSHA detects the condition during an inspection, OSHA will not use the report as evidence of willfulness. A timely, good faith response includes promptly taking diligent steps to correct the violative condition, while providing effective interim employee protection, as necessary.

OSHA will treat a voluntary self-audit that results in prompt corrective action of the nature described above and appropriate steps to prevent similar violations, as strong evidence of the employer's good faith with respect to the matters addressed. Good faith is one of the statutory factors that OSHA is directed to take into account in assessing penalties. 29 U.S.C. 666(j). Where OSHA finds good faith, OSHA's Field Inspection Reference Manual (the "FIRM") authorizes up to a 25 percent reduction in the penalty that otherwise would be assessed. The FIRM treats the presence of a comprehensive safety and health program as a primary indicator of good faith. A comprehensive safety and health program includes voluntary self-audits, but is broader in concept, covering additional elements. In this policy, OSHA has concluded that a voluntary self-audit/correction program is evidence of good faith. OSHA believes that the policy will provide appropriate positive recognition of the value of voluntary self-audits, while simultaneously enabling the Agency to enforce the provisions of the Occupational Safety and Health Act effectively.

V. FINAL POLICY CONCERNING THE OCCUPATIONAL SAFETY AND HEALTH ADMINISTRATION'S TREATMENT OF VOLUNTARY EMPLOYER SAFETY AND HEALTH SELF-AUDITS

A. Purpose

1. This policy statement describes how the Occupational Safety and Health Administration (OSHA) will treat voluntary self-audits in carrying out Agency civil enforcement activities. Voluntary self-audits, properly conducted, may discover conditions that violate the Occupational Safety and Health Act (Act) so that those conditions can be corrected promptly and similar violations prevented from occurring in the future. This policy statement is intended to provide appropriate, positive treatment that is in accord with the value voluntary self-audits have for employers' safety and health

compliance efforts, while also recognizing that access to relevant information is important to the Secretary of Labor's inspection and enforcement duties under the Act.

2. This policy statement sets forth factors that guide OSHA in exercising its informed discretion to request and use the information contained in employers' voluntary self-audit reports. All OSHA personnel applying this policy will receive instruction in order to ensure the consistent and appropriate application of the policy. The policy statement is not a final Agency action. It is intended only as general, internal OSHA guidance, and is to be applied flexibly, in light of all appropriate circumstances. It does not create any legal rights, duties, obligations, or defenses, implied or otherwise, for any party, or bind the Agency.

3. This policy statement has four main components:

    a. It explains that OSHA will refrain from routinely requesting reports of voluntary self-audits at the initiation of an enforcement inspection;

    b. It explains that OSHA will refrain from issuing a citation for a violative condition that an employer has discovered through a voluntary self-audit and has corrected prior to the initiation of an OSHA inspection (or a related accident, illness, or injury that triggers the inspection), if the employer also has taken appropriate steps to prevent the recurrence of the condition;

    c. It contains a safe-harbor provision under which, if an employer is responding in good faith to a violative condition identified in a voluntary self-audit report, and OSHA discovers the violation during an enforcement inspection, OSHA will not treat that portion of the report as evidence of willfulness;

    d. It describes how an employer's response to a voluntary self-audit may be considered evidence of good faith, qualifying the employer for a substantial civil penalty reduction, when OSHA determines a proposed penalty. See 29 U.S.C. 666(j). Under this section of the Act, a proposed penalty for an alleged violation is calculated giving due consideration to the "good faith" of the employer.

B. Definitions

1. "Self-Audit" means a systematic, documented, and objective review by or for an employer of its operations and practices related to meeting the requirements of the Act.

    a. "Systematic" means that the self-audit is part of a planned effort to prevent, identify, and correct workplace safety and health hazards. A systematic self-audit is designed by or for the employer to be appropriate to the scope of hazards it is aimed at discovering, and to provide an adequate basis for corrective action;

    b. "Documented" means that the findings of the self-audit are recorded contemporaneously and maintained by or for the employer;

    c. "Objective" means that the self-audit is conducted by or under the direction of an individual or group of individuals who are competent to identify workplace safety or health hazards, given the scope and complexity of the processes under review.

2. "Voluntary" means that the self-audit is not required by statute, rule, order, or settlement agreement. Voluntary self-audits may assess compliance with substantive legal requirements (**e.g.,** an audit to assess overall compliance with the general machine guarding requirement in 29 CFR 1910.212).

3. "Voluntary self-audit report" means the written information, analyses, conclusions, and recommendations resulting from a voluntary self-audit, but does not include matters required to be disclosed to OSHA by the records access rule, 29 CFR 1910.1020, or other rules.

4. "Good faith" response means an objectively reasonable, timely, and diligent effort to comply with the requirements of the Act and OSHA standards.

C. OSHA's Treatment of Voluntary Self-Audit Reports

1. No Routine Initial Request for Voluntary Self-Audit Reports
   a. OSHA will not routinely request voluntary self-audit reports at the initiation of an inspection. OSHA will not use such reports as a means of identifying hazards upon which to focus inspection activity.
   b. However, if the Agency has an independent basis to believe that a specific safety or health hazard warranting investigation exists, OSHA may exercise its authority to obtain the relevant portions of voluntary self-audit reports relating to the hazard.
   c. An employer voluntarily may provide OSHA with self-audit documentation and may be eligible to receive the benefits that are detailed in this policy.

2. No Citations for Violative Conditions Discovered During a Voluntary Self-Audit and Corrected Prior to an Inspection (or a Related Accident, Illness, or Injury That Triggers the Inspection)

   It is OSHA's current enforcement practice to refrain from issuing a citation for a violative condition that an employer has corrected prior to the initiation of an OSHA inspection (and prior to a related accident, illness, or injury that triggers the inspection), if the employer has taken appropriate steps to prevent a recurrence of the violative condition, even if the violative condition existed within the six month limitations period during which OSHA is authorized to issue citations. Consistent with this enforcement practice, OSHA will not issue a citation for a violative condition that an employer has discovered as a result of a voluntary self-audit, if the employer has corrected the violative condition prior to the initiation of an inspection (and prior to a related accident, illness, or injury that triggers the inspection) and has taken appropriate steps to prevent a recurrence of the violative condition that was discovered during the voluntary self-audit.

3. Safe Harbor – No Use of Voluntary Self-Audit Reports as Evidence of Willfulness

   A violation is considered willful if the employer has intentionally violated a requirement of the Act, shown reckless disregard for whether it was in violation of the Act, or demonstrated plain indifference to employee safety and health. Consistent with the prevailing law on willfulness, if an employer is responding in good faith to a violative condition discovered through a voluntary self-audit and OSHA detects the condition during an inspection, OSHA will not use the voluntary self-audit report as evidence that the violation is willful.

This policy is intended to apply when, through a voluntary self-audit, the employer learns that a violative condition exists and promptly takes diligent steps to correct the violative condition and bring itself into compliance, while providing effective interim employee protection, as necessary.

4. "Good Faith" Penalty Reduction

Under the Act, an employer's good faith normally reduces the amount of the penalty that otherwise would be assessed for a violation. 29 U.S.C. 666(j). OSHA's FIRM provides up to a 25 percent penalty reduction for employers who have implemented an effective safety and health program, including voluntary self-audits. OSHA will treat a voluntary self-audit that results in prompt action to correct violations found, in accordance with paragraph C.3. above, and appropriate steps to prevent similar violations, as strong evidence of an employer's good faith with respect to the matters covered by the voluntary self-audit. This policy does not apply to repeat violations.

D.   Federal Program Change

This policy statement describes a Federal OSHA Program change for which State adoption is not required; however, in the interest of national consistency, States are encouraged to adopt a similar policy regarding voluntary self-audits.

E.   Effective Date

This policy is effective July 28, 2000.

This document was prepared under the direction of Charles N. Jeffress, Assistant Secretary for Occupational Safety and Health, US Department of Labor, 200 Constitution Avenue, NW., Washington, DC 20210.

Signed at Washington, D.C. this 24th day of July, 2000.

Charles N. Jeffress,
Assistant Secretary of Labor.
[FR Doc. 00-19067 Filed 7-27-00; 8:45 am]
BILLING CODE 4510-26-P

Appendix B: OSHA Letter of Interpretation on Self-Audits

Reference: https://www.osha.gov/laws-regs/standardinterpretations/1996-09-11

September 11, 1996

Mr. Frank White
Vice President
Organization Resources Counselors, Inc.
1910 Sunderland Place, NW
Washington, D.C. 20036

Dear Mr. White:

Thank you for your letter to Secretary Reich concerning voluntary safety and health audits under the Occupational Safety and Health Act (the Act). Secretary Reich has asked me to respond. I appreciate Organization Resource Counselors' (ORC) interest in this issue. ORC's expertise in occupational safety and health issues is well established, and its views merit careful consideration.

Your letter takes issue with the Department of Labor's (the Department) practices regarding access to employer safety and health audits in Occupational Safety and Health Administration (OSHA) inspections. You state that the Department has not provided clear guidance as to the circumstances in which OSHA will seek disclosure of employer audits. You ask that the Department declare that it will not seek audit documents from an employer in conjunction with any inspection or investigation under the Act. You assert that with a few narrow exceptions, there are no federal requirements that an employer conduct a safety or heath audit. Your concern is that the possibility that audit results could be reviewed by the government may cause employers to refrain from conducting audits or may inhibit candor in the audit, **undermining** its usefulness. You explain employers may fear that audit reports would provide evidence of willful violations of the Act that, if disclosed to the government, could lead to assessment of large fines. Because audits are an important component of an effective safety and health program you believe it is important that the Department not create a disincentive to voluntary audits.

The Department shares your view that employer safety and health programs are fundamental to our effort to protect safety and health in the workplace, and that self-audits are an important part of an effective program. We strongly believe, however, that barring OSHA access to audit results would gravely impair the agency's ability to enforce the Act and to draw inappropriate distinctions between employers with effective and ineffective programs.

Such a policy is not necessary to encourage use of audits. Employers derive many benefits from effective safety and health programs that provide for audits, including reduced absenteeism, lower workers' compensation premiums and payments for medical treatment and disability, and favorable treatment from OSHA. Employers with effective programs have fewer and less serious hazards and thereby face reduced exposure to OSHA citations and penalties as a result. Moreover, employers found to have effective programs are eligible for limited scope inspections (in construction) and substantial penalty reductions for violations found in recognition of their good faith efforts. The concern that employers acting in good faith to respond to audit findings would be charged with willful violations rests on a misunderstanding of the relevant legal standards.

A BAN ON ACCESS TO EMPLOYER INFORMATION
WOULD IMPEDE ENFORCEMENT

1. ORC broadly defines a self-audit as any internal or external review of safety and health conditions or performance conducted by or on behalf of an employer" (p.2). This definition is broad enough to include almost any information that an employer has developed or obtained that is relevant to compliance with its OSH Act obligations; the definition would include information obtained or analyzes performed for the purpose of identifying hazards present in the workplace that are regulated by OSHA, determining the measures the employer will take to address the hazard and comply with its OSH Act obligations, and assessing the adequacy of those measures. A policy barring OSHA from access to this kind of information would gravely impair the agency's ability to enforce the Act.

The policy you suggest would allow OSHA to conduct inspections only by means such as visual observation of workplace conditions and the compliance officer's own physical monitoring efforts. Visual observation can be an effective technique for assessing compliance with requirements, particularly narrow specification requirements prescribing readily detectable physical measures within a reasonably small area. In many other situations, however, review of employer records and consideration of the employer's own analyzes and understanding of the sign are essential to an effective inspection.

This is particularly the case with requirements such as those that mandate the employer establish a program to address a hazard, or take measures to prepare for hazards that occur intermittently or change over time, or provide training to employees, or execute a continuing course of conduct or take appropriate protective measures based on its own assessment of hazards its workers face. As you know, there are many such requirements; new OSHA standards tend to be written in performance terms, rather than as narrow specification requirements.

For example, consider the general respiratory protection standard. That standard requires the employer to provide a respirator "when such equipment is necessary to protect the health of the employee"; the employer is to select respirators "which are applicable and suitable for the purpose intended" and is to anticipate and plan for "possible emergency and routine use." The employer is to establish a respiratory protective program and is to assure that respirators are "regularly cleaned and disinfected" and that the user of any respirator is "properly instructed in its selection, use and maintenance" (29 CFR 1910.134).

An effective inspection for compliance with these requirements must consider information the employer has compiled concerning workplace safety and health conditions and performance. The compliance officer may need to review information concerning the toxic substances employees are or may be exposed to over the course of their work and the sufficiency of engineering controls to limit the exposure. The compliance officer may also need to review records concerning the system the employer has established for maintenance of respirators, and the steps the employer has taken to train respirator users. Placing employer information of this kind, off limits would imperil the credibility of the inspection.

2. The premise for ORC's position is that the review of safety and health conditions and performance in the workplace is purely optional with the employer. It is argued that OSHA must not seek access to the information in these reviews, because if the agency does, employers will stop conducting them. ORC's position against disclosure, as we understand it, does not include audits that are required by OSHA standards. ORC asserts, however, that with a few narrow exceptions, there are no federal requirements for audits.

This position, we respectfully suggest, misapprehends the scope and degree of existing requirements that employers conduct audits as ORC defines that term. Employers in the construction industry are subject to a comprehensive audit requirement. They must institute a safety and health program that provides for frequent and regular inspections of job sites by competent persons to assure compliance with the OSHA construction standards (29 CFR 1926.20(b)). There is no comprehensive requirement for general industry employers to conduct audits, but there are many audit requirements in individual standards, including fundamental generic standards. The confined spaces standard, for example, requires employers to evaluate whether their workplace includes confined spaces, to establish a written program if employees will be required to enter confined spaces, to monitor and record conditions for each entry, and to review annually the program entries conducted and make necessary modifications (29 CFR 1910.146). The lockout/tagout standard requires employers to develop energy control procedures for servicing and maintenance and to conduct periodic inspections to ensure that the procedures and the requirements of the standard are followed (29 CFR 1910.147(c)). The process safety management standard requires comprehensive process hazard analyzes, mechanical integrity inspections, incident investigations, and compliance audits (29 CFR 1910.119). The hazardous waste standard includes similar requirements (29 CFR 1910.120).

The general personal protective equipment (PPE) standard requires employers to conduct an assessment of the hazards employees are likely to be exposed to, to select appropriate PPE based on the assessment, to train employees, and to assure that employees have understood the training (29 CFR 1910.132). There is to be daily inspection and periodic testing of electrical PPE (29 CFR 1910.137(b)). The general respirator standard requires employers to establish written procedures governing selection and use, to select respirators based on a hazard assessment, to maintain appropriate surveillance of work area conditions and degree of employee exposure or stress, and to conduct regular inspection and evaluation to determine the continued effectiveness of the program, including inspections of all respirators before and after each use (29 CFR 1910.134(b), (f)). Many narrower hazard-specific standards also require employers to assess workplace conditions or inspect for compliance with requirements. See, e.g., 29 CFR 1910.272(g)(1) (grain handling).

Health standards issued by OSHA under Section 6(b) of the Act commonly contain similar provisions. The lead standard, for example, requires employers to conduct monitoring to evaluate employee exposures to airborne lead, to develop a written compliance plan to reduce exposures to the permissible level and to revise and update the plan semi-annually (29 CFR 1910.1025(d), (e)). See also 29 CFR 1910.95(c) (hearing conservation).

Where obligations such as these are involved, an inspection would be impossible if OSHA were barred from access to employer reviews of workplace conditions or performance, because the obligation is to conduct such a review. The scope and number of such requirements suggests that it would be no easy task to disentangle those parts of a comprehensive review that are voluntary from those that are not. But even where an OSHA standard includes no explicit obligation to review workplace conditions or performance to facilitate compliance with the standard, the employer is obligated to comply with the standard itself and the steps the employer has taken to assure compliance are highly relevant to enforcement of the Act. The courts have stressed that the OSH Act does not impose absolute liability on employers for non-compliance with a standard, but that it does require diligent efforts to comply, see Horne Plumbing & Heating Company v. Occupational-Safety and Health Review Commission 528 F.2d 564 (1976).

The OSH Review Commission has held that to prove a violation of the Act, the Secretary must show not only that a violative condition exists, but that the employer had actual or constructive knowledge of the condition, see CF&T Available Concrete Pumping, 15 BNA OSHC 2195, 1991-93 CCH OSHD **29,945 (No. 90-329,1993). The Secretary must show that the employer knew, or with the exercise of reasonable diligence could have known, of the violative condition, Ibid. The state of the employer's knowledge and the diligence of the methods it has employed to find and prevent violations are therefore of central importance to investigation and enforcement of the requirements of the Act. Employer reviews of safety and health conditions or practices that are relevant to compliance with a standard have a direct bearing on whether the employer has met its obligations. A policy barring OSHA from access to information of this kind would undermine enforcement of the Act.

Access to information of this kind is also essential to classification of violations and calculation of penalties. Under Section 17(j) of the Act, a penalty must be based in part on the employer's good faith. OSHA has interpreted good faith as referring to the employer's establishment of an effective safety and health program, which includes audits. Existing guidelines in the Field Inspection Reference Manual (FIRM) authorize a reduction of 25% in the penalty for employers who have implemented such programs. See FIRM page IV-14 at C.2.L(5)(b). As discussed below, OSHA is in the process of establishing initiatives, referred to as the New OSHA, that will substantially increase the discount for superior and outstanding programs.

In short, the policy you suggest would severely impair OSHA's ability to enforce the Act. The policy would undermine the agency's ability to inquire into the existence of violative conditions, to establish employer knowledge, to classify violations found, and to assess penalties.

THE NEW OSHA

Several of the initiatives announced in the May 1995 National Performance Review report, The New OSHA, depend on the agency's acquiring a thorough understanding of the employer's worksite safety and health program, including the employer's evaluation of safety and health hazards present and the steps the employer takes to address them. The policy ORC proposes would preclude OSHA's obtaining this information.

As you know, the central concept of the New OSHA initiatives is that OSHA should emphasize the state of the employer's safety and health program, rather than simply inspecting for compliance with individual standards. Although many employers have a safety and health program the programs vary dramatically in scope and effectiveness. OSHA has prepared, with help from ORC and others, the Program Evaluation Profile ("PEP"), which is presently undergoing field testing. The PEP analyzes employer programs on fifteen factors, and assigns a numerical score for each factor. Some of the important factors include comprehensive worksite survey and hazard analysis, regular site inspection, employee hazard reporting system and response, accident and "near-miss" investigation, and injury and illness data analysis, all of which require an audit as ORC uses that term. OSHA must be able to review information concerning the employer's performance on these factors for the New OSHA initiatives to work.

OSHA's intention is that employers who score well on the PEP will obtain important benefits, including large reductions in penalties for serious violations, and elimination of penalties for other-than-serious violations. The New OSHA demonstrates an alternative means of recognizing and rewarding employer safety and health efforts that is superior or outstanding. The audit access ban ORC proposes would shield all programs, good, bad, or indifferent from inquiry. Even records of known hazardous conditions would be off limits to OSHA. The audit access ban would prevent OSHA from understanding the state of the employer's efforts and from treating employers with superior or outstanding programs differently from employers with ineffective, developmental or basic programs. The new OSHA approach on the other hand, allows a detailed assessment of employers' health and safety performance. Employers who have done a good job receive favorable treatment, while poor performance can be identified and remedied.

AN ACCESS BAN IS UNNECESSARY

An employer derives many significant benefits from an effective safety and health program that provides for self-audits. These benefits arise both within and outside the ambit of the OSH Act. Employers who conduct effective self-audits receive substantial advantages in OSH Act inspections compared with those who don't. We therefore do not agree that an audit access ban is necessary to induce employers to conduct audits.

An effective self-audit procedure, is part of a comprehensive safety and health program, should reduce employee injuries and illnesses, saving the employer costs resulting from absenteeism, workers' compensation and other insurance payments. An effective program may help reduce employee turnover and improve productivity. In terms of the OSH Act, the principal consequence of an effective audit program is a reduction in the number and severity of hazards, leading to a corresponding reduction in citations and penalties in the event of an inspection. A conscientious program should be particularly effective in eliminating high gravity serious, willful, repeated, and failure to abate violations, which carry by far the heaviest penalties.

In view of all these benefits, we find it difficult to believe that companies will stop implementing comprehensive safety and health programs or conducting audits if OSHA retains its present policy. Moreover, even if OSHA were to adopt a policy against inquiry into audit information, that policy would not make such information truly confidential. Occupational health audits would generally be subject to the records access rule, which guarantees a right of access to employees (29 CFR 1910.1020). If employees are represented by a union, employer information about workplace safety and health must be disclosed upon request to the union, as an incident to the company's duty to bargain in good faith about safety and health issues, see NLRB v. American National Can Company, Foster-Forbes Glass Division, 924 F.2d 518, 524 (4th Cir. 1991).

Finally, such information would apparently not be protected from disclosure in private tort litigation. The courts have generally rejected claims to withhold information of this kind in discovery under a "self-evaluative privilege." The Ninth Circuit addressed the issue in Dowling v. Amedcan Hawaii Cruises, 971 F.2d 423 (9th Cir. 1992). The court stated that voluntary audits are rarely curtailed because they may be subject to discovery in litigation. Noting that companies typically conduct such audits to avoid litigation resulting from unsafe working conditions, the court found ironic the claim that such candid assessments will be inhibited by the fear that they could later be used as a weapon in the hypothetical litigation they are intended to prevent.

In short, employers who conduct effective audits derive many advantages, including advantages in the event of an OSHA inspection, from the practice. They have no need of the shield against access that you suggest which in any event could not make the audits truly confidential. We are nonetheless concerned by your statements that some employers perceive a disincentive to perform self-audits from OSHA's policies. In order to address this perception, it may help to describe relevant elements of OSHA's citation policy and the case law under the Act.

The purpose of self-audits is to find hazardous conditions and remedy them. If a self-audit discloses a condition that is a violation of the OSH Act presumably the employer will take action to correct the problem. In the event the employer permanently remedies the condition before an OSHA inspection takes place (and before the occurrence of an accident or other event triggering an inspection), including taking appropriate steps to prevent a recurrence of the violation, OSHA's practice is not to issue a citation, even though the violation may have existed within the six month statute of limitations period. If the violation has been permanently corrected on the employer's own initiative without the need for action or intervention by OSHA, the agency sees no need to spend its own limited enforcement resources addressing the problem. Further, as noted, evidence that the employer is finding and fixing problems on its own will weigh heavily in the employer's favor for purposes of good faith.

If, on the other hand, an employer has identified a violative condition in an audit and has failed to abate it, and the OSHA inspection finds the violation, a citation may issue. Even here, however, good faith efforts made in response to the audit will benefit the employer. If the employer has responded promptly to the audit and believes in good faith, although erroneously, that it has resolved the problem and come into compliance with the OSHA standard, that would tend to negate willfulness. The Review Commission has frequently held that an employer's reasonable good faith belief that its actions comply with a standard is inconsistent with willfulness, although the actions were in fact incomplete and do not fully remedy the hazard, see Calang Corp., 14 BNA OSHC 1789, 1987-90 CCH OSHD **29,080 (No. 85-319, 1990). In short, the concerns you have expressed that conscientious employers who conduct audits would expose themselves to willful citations are based on a misunderstanding of the case law and the Secretary's citation policy.

Of course, if the employer has simply ignored the audit finding of a hazardous condition, the employer will get no credit for the audit. Such an employer could benefit from a policy barring access to audit information. We see no reason, however, for rewarding an audit program that takes no action to remedy identified hazards. We expect however, that there are few employers in this category. Responsible employers who react conscientiously to audit findings will benefit themselves and their workers.

We would be pleased to meet with you to discuss the issues addressed in this letter, should you consider such a meeting useful.

Sincerely,
Joseph A. Dear
Assistant Secretary

Appendix C: OSHA Compliance Checklist

Reference: https://www.osha.gov/sites/default/files/2018-12/fy11_sh-22318-11_Mod_1_OSHAinspectionchecklist.pdf

Checklist for OSHA Compliance Inspections

Farm Name: _____ Date: _____ Time: _____

Name of person using checklist : _____

Inspectors have statutory authority to:	Other provisions

○ Arrive unannounced
○ Enter without delay and at reasonable times
○ Inspect and investigate the workplace:
  ▪ during regular working hours
  ▪ at other reasonable times
  ▪ within reasonable limits and in a reasonable manner
	○ Question privately any employee or employer
	○ **Other provisions**
	• Confidentiality–Names of complainants can be kept confidential
	• Participation in inspection

ARRIVAL / OPENING CONFERENCE

Credentials: A person states their intention to conduct an occupational safety inspection of your farm. Ask this person for their credentials.

• Federal Credentials   Yes ☐   No ☐

Name of Compliance Safety and Health Officer (CSHO) _____

If credentials are acceptable, proceed to next item. To verify credentials, call area OSHA Director. See OSHA contacts on next page.

Purpose and Scope of Inspection

What is the impetus for the inspection?   ☐ Employee complaint   ☐ Program Inspection   ☐ Referral ☐ Other (describe) _____

NOTE: Ask to see employee complaint or referral. Attach photocopy to your final notes. Inspector's failure to provide details of employee complaint (other than identification of employee) may be cause for appeal.

Contact phone number(s) of additional farm management team or individual responsible for safety program to be involved in the inspection process.

_____

Ask the inspector what is the purpose and intended scope of the inspection (provide summary).

Employee Participation

With above information on purpose and scope of inspection, consult with the CSHO as to appropriate employee representation. If necessary, contact employee representative to attend the inspection.
Summarize the agreement regarding employee participation in the inspection.

Miscellaneous Items

• Plan and state your proposed route of inspection that will cover the purpose and scope of inspection.
• Gather up notebook, checklist, camera, two-way radio or cell-phone, and list of farm management team.

ON-SITE INSPECTION

Records and written programs: Examples of items you should be prepared to show.

• OSHA 300 logs	• Confined space programs
• HazCom program, MSDS records	• Lockout/tagout
• Employee training records	• Respiratory protection program

Notes, photos and measurements

- Notes—Names of people participating in on-site inspection, times, places visited, CSHO's comments, names of people spoken to, your observations, etc.
- Corrections—Where possible, immediately correct violations pointed out by the CSHO. Make a note and take a photo of your actions.
- Photos—If the CSHO takes a photo, you take the same photo. Ask CSHO why the photo was taken.
- Measurements—Take any measurement taken by CSHO, or ask for copy or reading.

CONCLUDING THE INSPECTION

Closing conference: At the conclusion of the on-site inspection, ask for a closing conference.

At the closing conference, allow the CSHO to address their findings. Take careful notes on their statements at the closing conference. If you are less than completely clear about their findings, restate your understanding of their findings to the CSHO for agreement.

If they have not addressed the following issues, be sure to ask for answers.

- What are the alleged violations?
- What are the CSHO's next steps in the process?
- Will there be further on-site inspection prior to issuance of any citations or 'decision not to issue'?
- When can your farm expect to receive any 'decision not to issue' or citations?

After the CSHO departs

- Formalize your notes, photos and measurements.

Wisconsin OSHA Area Office Contacts

Appleton Area Office
(920) 734-4521

Madison Area Office
(608) 441-5388

Eau Claire Area Office
(715) 832-9019

Milwaukee Area Office
(414) 297-3315

Checklist for OSHA Compliance Inspections, February 2012.
C.A. Skjolaas, Agricultural Safety Specialist, UW-Madison/Extension Center for Agricultural Safety and Health
Adapted from UW-River Falls Checklist for Department of Commerce Inspections developed by Constance Smith, Director of Risk Management, UW-River Falls.
This material was produced under grant number SH-22318-11-60-F-55 from the Occupational Safety and Health Administration, U.S. Department of Labor. It does not necessarily reflect the views or policies of the U.S. Department of Labor, nor does mention of trade names, commercial products, or organizations imply endorsement by the U.S. Government.

Index

Note: Locators in *italics* represent figures and **bold** indicate tables in the text.

A

Access control, 213
Accountability, 154–156, 207, 218
Acme distribution, 102
Acme manufacturing, 117
Adult education; *see also* Ongoing avenues of
      education
  compliance training, 57–58
  face-to-face sessions, 57
  generations, 56
  principles of, 56
  skill development, 59
Agility, 65
Airline tickets, 169
Alarm activations, 218–219
The American National Standards Institute
      (ANSI), 20, 33–34
American Sign Language (ASL), 91
The American Society of Safety Professionals
      (ASSP), 33–34, 50
Americans with Disabilities Act, 160
Analytical skills, 162
Andragogy, 164
Annual audits, 187, 222
ANSI/ASSP Z10, 33–34, 37
Audit, 51–52; *see also* Closing conference
  auditor's activities, 104
  data entry, 109
  documentation review, 107–108
  facility management, 8
  and inspections, 7
  measurement outcomes, 8
  opening conference, 104–106
  organizational levels, 7
  performance of systems, 3
  pre-audit phase, 104
  regional levels, 8
  tools, 8–9
  training, 109
Audit content, 199–200
Audit document, 184
  distributed, 114–115
  follow-up, 115–116
  information, 112–113
  proofread, 113–114
Audit format, 200
Audit frequency, 182–183

Auditing goals, 206
Auditors
  calibration, 158
  complexity of operation, 100
  regulation and company policy review, 101
  scope of audit, 100
  selection and training, 187
  selection process, 160
  size of facility, 100
  time allotted for audit, 100
  travel, 101–102
Auditor training
  andragogy, 164
  challenges, 162–163
  experiential learning, 164
  storytelling, 164
  training environment and content,
      164–166
  workplace, 163
  in world, 163–164
Auditor work area, 88
Auditor workspace, 170–171
Audit programs; *see also* Auditors
  auditor selection and training, 97
  documentation review, 100
  facility management schedules, 98
  facility personnel, 100
  holidays and vacations, 98
  logistical considerations, 99
  ongoing communications, 99
  periodic communication, 99
  personnel needs, 99–100
  pre-audit phase, 97
  production schedules, 97–98
  schedule of, 98–99
  time between, 98
  unannounced, 99
Audit report, 185
Audits and inspections, 44–45
Audit scoring, 110
  compliance/non-compliance, 156
  methodologies, 156
  scaled scoring, 156–157
Audit team location
  confidentiality, 172
  technology, 171–172
  workspace, 172
Audit title, 132–134

Printed in the United States
by Baker & Taylor Publisher Services

Printed in the United States
by Baker & Taylor Publisher Services